United We Sense: How Cars Talk to Each Other for Enhanced Perception

Parkar

First Printing, 2024

Content

1 Introduction

Nowadays, the development of the automotive industry focuses on safety and reducing traffic accidents. Many types of research work on the robot system, intelligent transportation system (ITS) and advanced driver assistance systems (ADAS). One of the primary fundamental processes of these systems is acquiring information from the environment. The correctness and reliability of data are essential for the real-time task of the system. Recent system configurations in automated cars increase the number of sensors to improve and extend the cars spatial detection limits. Future automated driving use cases will depend on high range perception (for example due to high-velocity situations on German highways). Single sensor systems cannot reliably reach detection ranges beyond 200m. That is why cooperative perception will be introduced to extend the sensorial perception beyond current limits. Cooperative means that multiple cars (agents) will share their perception in real-time and thus improve their scene understanding.

At the beginning of this report, we give an overview of machine learning approach and semantic segmentation method for understanding why we need them for online perception. In the next chapter, we describe Bonnet [1], one of the first frameworks for semantic segmentation, and how we can improve this framework. The development and implementation of the online perception system are given in the next part of the report. Furthermore, we evaluate the performance of our development system to clarify whether the system meets our requirement or not. In the end, there are about the conclusion and the future of the research.

2 Machine Learning

Machine Learning (ML) method, which has attracted increasing attention in the last decade, is a subfield of Artificial Intelligence (AI) domain that provides systems the ability to improve by learning from experience without being specifically programming. At the beginning of this chapter, we give an overview of ML, what are the traditional methods and modern ones. Later on, we give a brief look at the neural network model, which is widely used on modern ML method such as deep learning. After that, we go into detail of the deep learning method in term of definition and example use case.

2.1 Overview

The first definition of ML is often attributed to Arthur Samuel in 1959 [2], who link the term "machine learning" with the definition "without being explicitly programmed" [2]. Later on, the more detail of ML is given in 1997 by Tom Mitchell [3], he describes the program that can learn from experience E concerning some class of task T with performance measure P, if its performance on T, as measured by P, improves with experience E [3]. In other words, ML extracts patterns from given data to further make predictions on future data. Machine learning algorithms are often categorized into three different categories: supervised machine learning, unsupervised machine learning, and reinforcement learning.

- **Supervised machine learning**: The program starts to analyze a known training dataset (contains both the inputs and the desired outputs) and apply the learning algorithm to produce an inferred function, which is used later to make correct predictions for new input data [4].

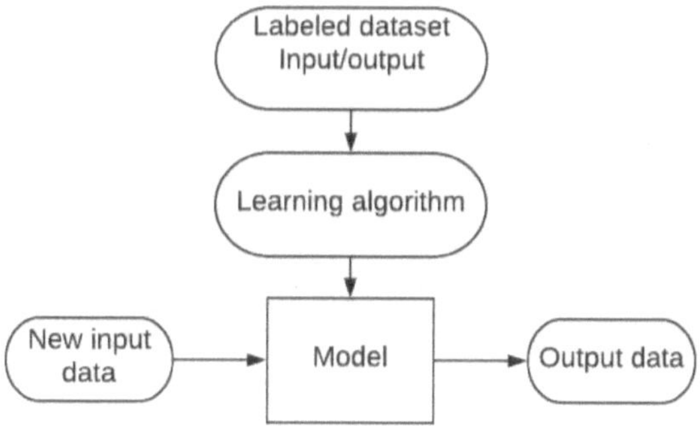

Figure 2.1 Supervised machine learning

- **Unsupervised machine learning**: We use this method when the information used to train is neither classified nor labeled. The program is given a bunch of unlabeled data and infer function to describe a hidden structure   (patterns) and relationships [5].

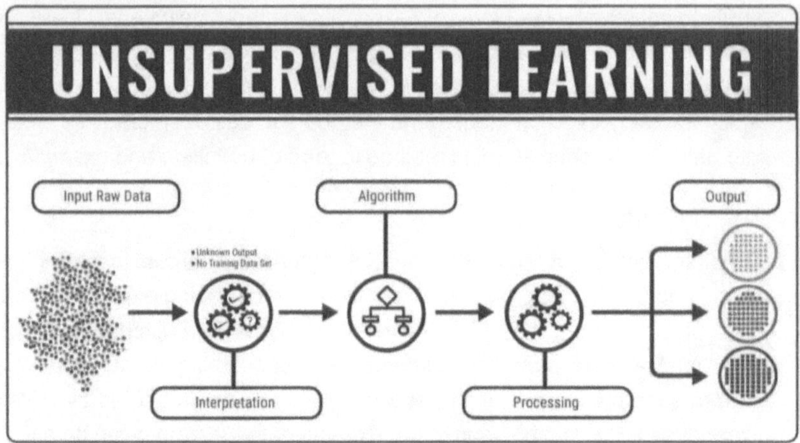

Figure 2.2 Unsupervised machine learning [6]

- **Reinforcement learning**: Reinforcement learning is a learning method that the program explores its dynamic environment by producing actions and try to maximize some notion of cumulative reward. In this method, the program requires the knowledge of estimation of the correctness of the answer [7].

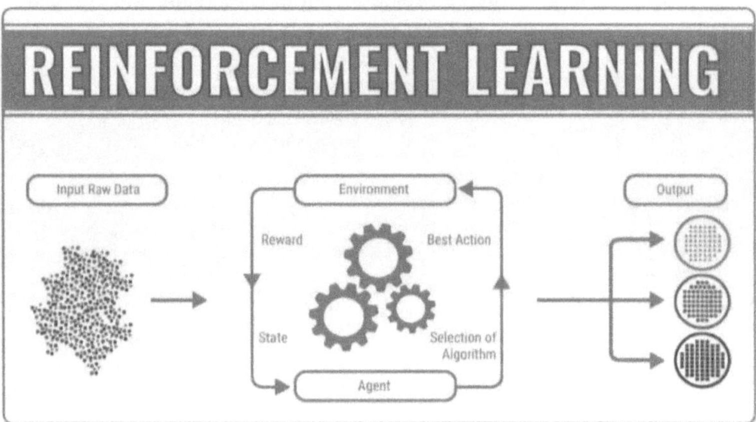

Figure 2.3 Reinforcement learning [6]

2.2 Neural Network

Artificial neural networks (ANN) are one of the leading frameworks for many different machine learning algorithms. They are inspired by how human learning with our brain. In the early 1940s, the history of ANN began with the pioneering work of Warren McCulloch and Walter Pitts [8]. The ANN model is developed based on biological neural networks (Figure 2.4).

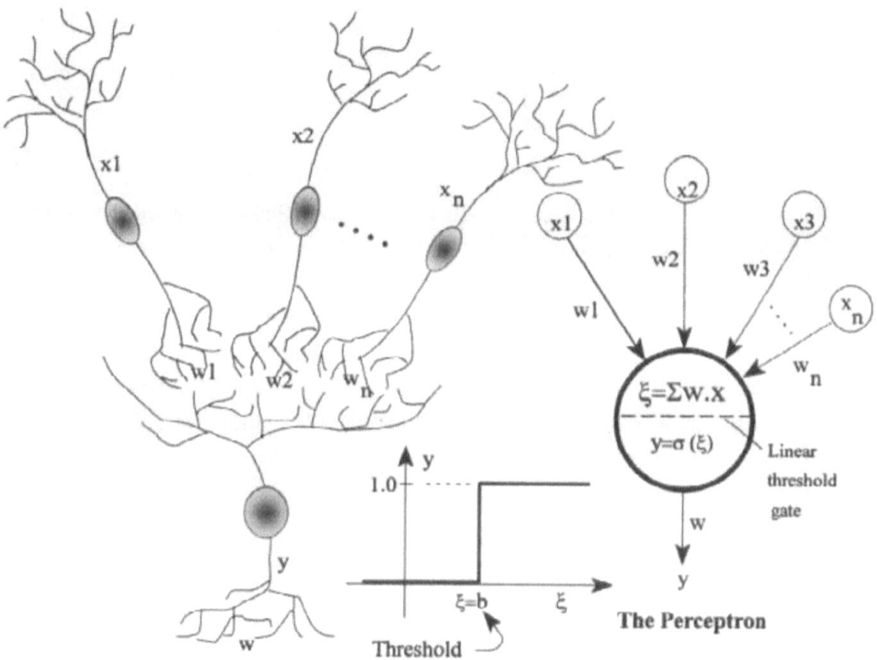

Figure 2.4 Signal interaction from n neurons and analogy to signal summing in an artificial neuron comprising the single layer perceptron [9]

One definition for an ANN was introduced first in [10], where Maureen Caudill stated that "a neural network is a computing system made up of a number of simple, highly interconnected processing elements, which process information by their dynamic state response to external inputs [10]".

In ANN, they are input, output layers, and hidden layers. The hidden layer procedure an intermediate result that the output layer can use. Each layer consists of neurons with an activation function. The example of a basic 3-layered is shown in Figure 2.5.

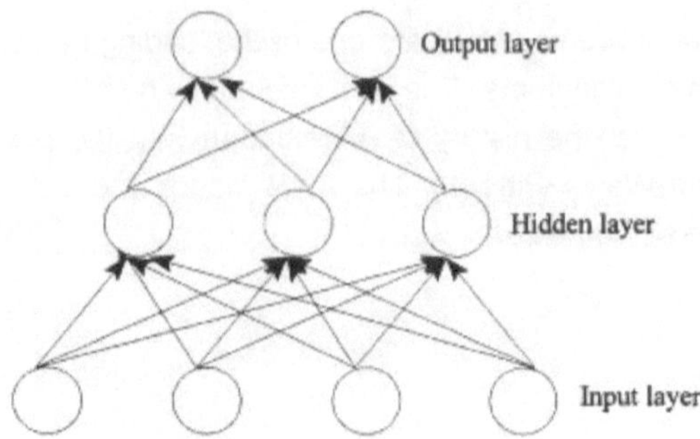

Figure 2.5 Three layer neural network [9]

Even though ANN have been introduced since the 1940s, it only becomes well-known in recent decades, because of the limitation of the simple perceptron [11]. Since the 1980s, the ANN became a significant part of the AI field, when the researcher introduced "backpropagation" technique [12]. The "backpropagation" technique allows neural networks to improve their hidden layers when the outcome does not equal the desired output. In the next section, we give fundamentals about Deep Neural Network (DNN), which was developed from traditional ANN. Moreover, Convolution Neural Network, one of the most popular architectures in Deep Neural Network, and the learning process, a crucial part of ML, are described.

2.3   Deep Neural Network

The artificial neural network with many hidden layers is called a deep neural network (DNN). DNNs recently has seen massive success in tasks such as image segmentation [13] [14] [15], image recognition [16], speech recognition [17] [18] and handwritten recognition [19] [20]. In DNNs, the network goes through all the layers to determine the probability outputs and are able to learn low level and high-level representations from given data [21]. This property makes the DNN can learn without any hand-craft feature, which makes the algorithm save a significant amount of human effort. Figure 2.6 shows an example of learned features, from low to high-level.

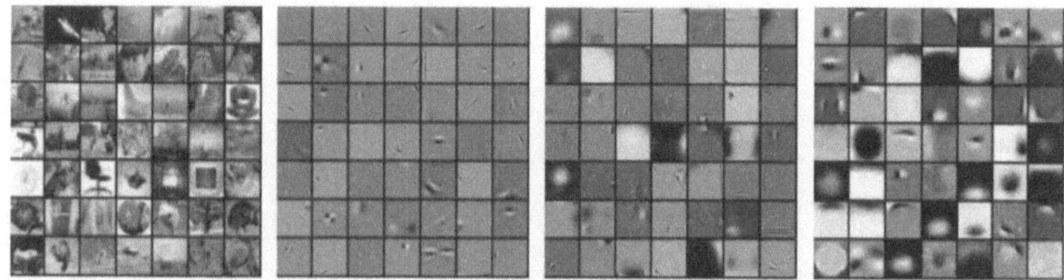

a.Input images   b.Low-level features   c.High-level features

Figure 2.6 Low-level and high-level representation of images [22]

With the help of many hidden layers, DNN can capture low-level features as corners and edges in low layers and high-level features in higher layers, which lead to the capability of modeling complex composition of object classes. DNNs can model not only linear relationships but also complex non-linear relationships. There are several architectures from DNNs such as Convolutional Neural Networks (CNN), Recurrent Neural Networks and Recursive Neural Networks. Despite the high performance of DNN, there is still some significant limitation of DNN such as the requirement for a huge amount of labeled data and computation resource (CPU and GPU). These limitations of DNN are the primary effect on the high cost of development of a program using DNN.

In the next part of this chapter, we will go to the detail of CNN, which is one of the first DNN architectures and also the architect that we use for our online perception system.

2.3.1 Convolutional Neural Network

In this chapter, some fundamentals about Convolution Neural Network and its structure will be described.

In DNN, Convolutional neural networks are used mainly for images, video, and natural language recognition. Since 2012, after the winning of Alex Krizhevsky in ImageNet [23] by dropping the classification error record from 26% to 15%, CNN has probably become the most useful developments in the field of computer vision.

The idea behind CNN base on the property of images that nearby pixels are more correlated than distant ones. CNN is similar to native neural networks. It is the combination of neurons with learnable weights and biases, but instead of all the neurons (nodes) being fully connected, each node only connected to a subset of the input image. It helps the forward functions more efficient and vastly reduces the number of parameters in the network. One more advantage of CNN over traditional neural networks is that the layers of neurons have arranged in width, height, and depth. An example architecture of CNN, which one of the first CNNs which was used to recognize digit [24] is in Figure 2.7 below.

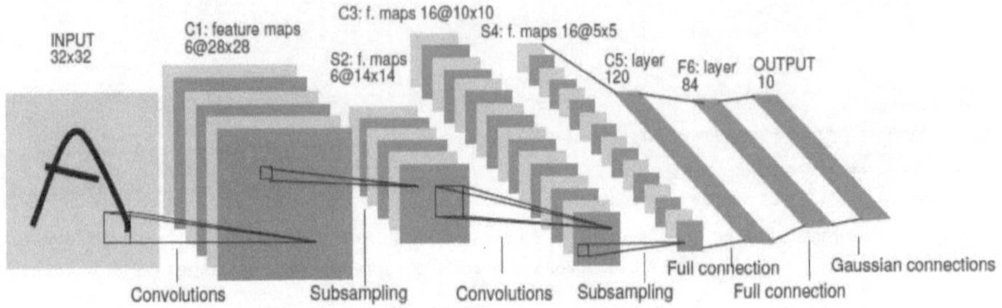

Figure 2.7 Example architecture of CNN [24]

Technically, a simple CNN is a sequence of layers and several layers of convolution associated with nonlinear activation functions to produce abstract/ higher-level information for subsequent layers. CNN architectures contain four main types of layers to build [25]:

- Convolutional layer
- Pooling layer
- Activation layer
- Fully connected layer

Full CNN architecture is the combination of these layers. The arrangement on the number and order among these layers will create different patterns suitable to use for different purposes. A typical stage of a CNN, which convolution, activation (non-linearity) and pooling layers are applied sequentially is shown in Figure 2.8.

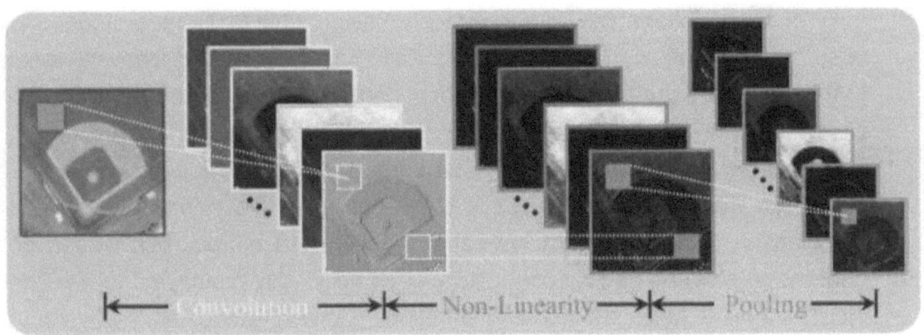

Figure 2.8 Illustration of a typical stage of a CNN [26]

2.3.1.1 *Convolutional layer*

Convolutional Layer is the first layer in a CNN. The convolutional layers are the main building blocks of a Convolutional Neural Network. They apply a convolution operation (a filter or a kernel) to the input. The filter is also an array of numbers which are called weights or parameters. After convolving the input data with the filter, we will get an output which is called a feature map or activation map. If the CNN is applied to images, then the convolution is typically implemented on a 3-dimensional input which results in a 2-dimensional output. Each layer contains N filters, and each filter produces results in an activation map. These results are stacked with each other over all dimensions to achieve the output of a layer. Figure 2.9 shows an example of a convolutional layer

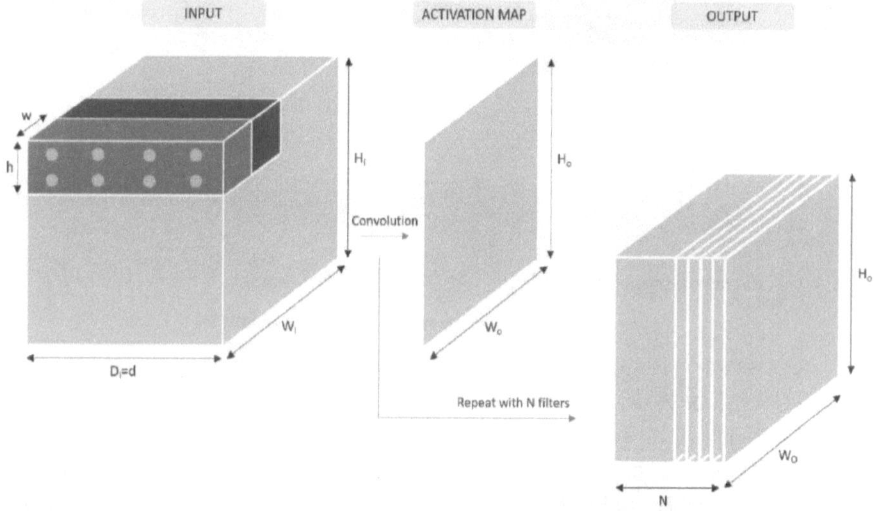

Figure 2.9 An example of a convolutional layer [27]

We can calculate the output volume based on the dimensions of filter and input image.

$$H_0 = \frac{H_i - f + 2p}{s + 1}$$

$$W_0 = \frac{W_i - f + 2p}{s + 1}$$

$$N = D_0$$

In the equation above, we assume that the filter is square ($h = w = f$). The intervals the filter moves in each spatial dimension is the stride s of a filter, and padding p is the number of pixels added at the outer edges of the input [7]. One more important thing about the convolution layer we have to remember is that the dimension of filter need not always be the same in the different layer. In the later chapter, we will see that changing the size of the filter in different layers help us to improve the performance of the CNN.

2.3.1.2 Activation layer

This layer is usually applied immediately after the Convolutional layer. It takes the outputs from a convolution layer into functions, which are non-linearity functions. There are many activation functions such as Tanh, Sigmoid function, and Rectified Linear Unit (ReLU).

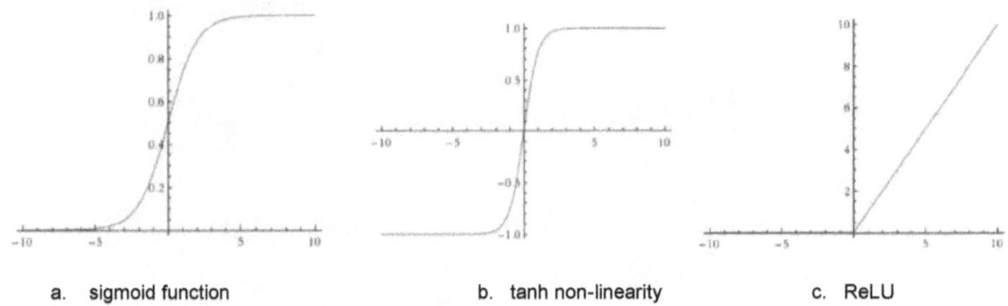

a. sigmoid function	b. tanh non-linearity	c. ReLU

Figure 2.10 Sample activation functions [25]

ReLU function is the most broadly utilized activation function in neural networks today. Rectified Linear Units (ReLU) layer applies a nonlinear function to increase nonlinear properties in a system that has been computing as linear properties during convolutional layers without affecting the reception fields of the convolutional layer. The responsibility of ReLU layer is converting all negative result from the Convolutional layer to 0.

$$f_x = max(0, x)$$

The main reason for using ReLU over other nonlinear function is the speed of training. In [23], researchers show that DNN with ReLUs trains several times faster than an equivalent network with *tanh* neurons.

2.3.1.3   Pooling layer

Pooling layer's operation has some similarity to the operation of Convolutional layers. Such as in convolution layer, pooling layer has a sliding window called a pooling window, which slides through each value of the input data matrix (usually the feature map). When a window slides over the input data matrix, only one value, which is considered the value representing the data information in that region (the sample value), is retained. The outcome of the pooling process is down-sampling, which means to reduce the dimensions of the feature space. The standard methods in the Pooling layer are [28]:

- Max Pooling: taking the maximum value.
- Min Pooling: taking the minimum value.
- Average Pooling: taking the average value.
- L2 Pooling: taking the square root of the total of the squares

By discarding redundant information, Pooling layers help CNN to decrease the computational cost. Moreover, these operations provide a form of robustness to CNN [29]. The illustration of pooling layers is shown in Figure 2.11.

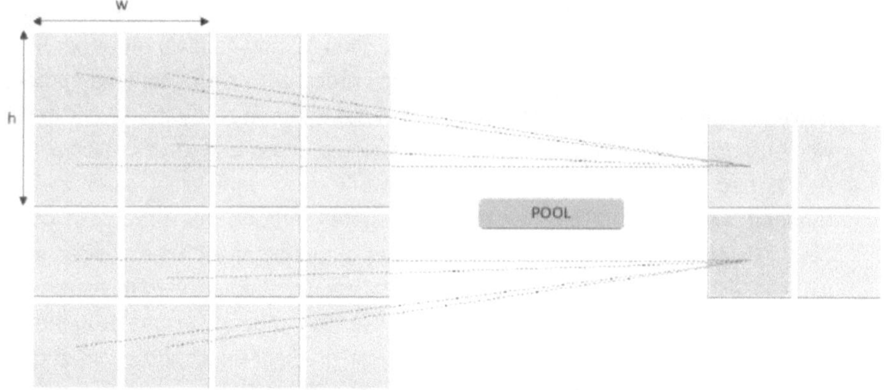

Figure 2.11 Illustration of a 2x2 pooling layer [27]

2.3.1.4   Fully Connected Layer

When higher-level features are identified by the previous layer (pooling and convolution), Fully Connected (FC) layer takes a look at what high level features most firmly correlate to a particular class and has specific weights. In this layer, all neurons are connected to the whole input volume so that when the users process the results between the weights and the previous layer, they can decide which features match with which class by the probabilities. For this reason, FC layers are normally the last layers of CNN. This layer takes an input volume and outputs an N-dimensional vector where N is the total of classes of our dataset.

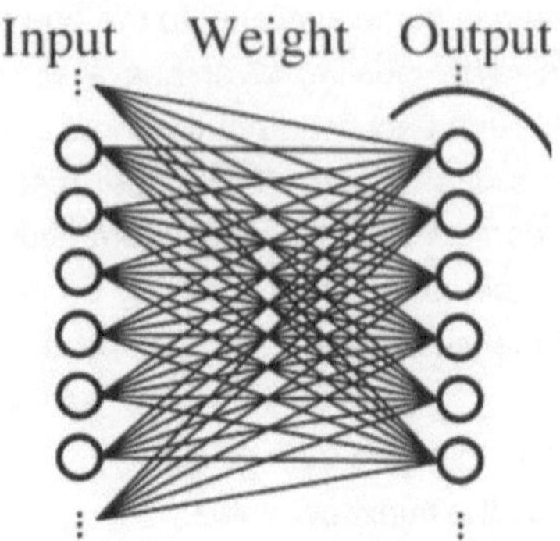

Figure 2.12 Example of FC layer [30]

2.3.2 *Learning Process*

In this previous section, we have fundamental knowledge about CNN. We know that the parameters (weight and bias) of neurons in CNN determine the characteristics and the performance of the network. The process of finding values of parameters is the learning process of CNN, which is the most genuine and crucial part of not only CNN but also Deep Learning. The learning process is an iterative process of forwardpropagation and backpropagation of the information by the layers of neurons with error/loss optimization. The overall learning process is given as follow.

At the beginning of the process, the network starts with initial parameters (often random). An example of input data propagates through the network to obtain the network's prediction. After that, the predicted output along with the desired output go to the loss function for comparing the data and computation error. At this step, we apply the backpropagation in order to propagate this loss to each parameter in the NN and update the parameters of the network with the gradient descent technique to reduce the loss and obtain a better model. The process keeps training until the model meets specific requirements (time out, number of iterations or desired loss). Figure 2.13 below shows the basics of the learning process. In the next part of this section, we are going into details of the important specification of learning processes.

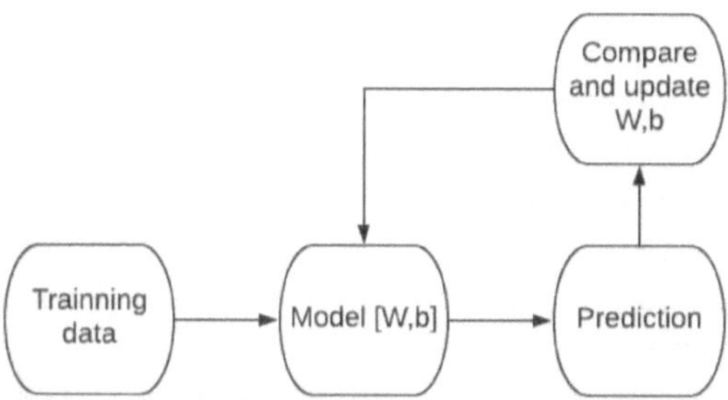

Figure 2.13 Basic Learning Process

2.3.2.1 Loss function

We need the loss function to evaluate how well the CNN model in terms of being able to predict the expected output. The smaller the result of the loss function, the better the model. The ideal model has the result of loss function equal to 0. There are two main loss function: mean-square and log-likelihood function. The choice of loss function depends on which problem we work with, the mean-square error function mainly uses for regression task, while the log-likelihood mainly use for dealing with classification problems.

$$l_i(\theta) = \frac{1}{2}|p_i - y_i|^2$$

$$l_i(\theta) = -y_i\log(p_i)$$

With l_i is individual losses, y_i is desired output and p_i is predicted output.

Because mean-square and log-likelihood function are linear, we only care about the individual losses. The total loss is sum up, which we need to minimize. The simplest method to do this is using stochastic gradient descent technique.

2.3.2.2 Stochastic Gradient Descent

Stochastic Gradient Descent (SGD) is an iterative method, which is firstly developed in 1951 [31]. This method is used for optimizing function, which in our case is loss function in the learning process. We apply SGD iteratively until the network converges. The function below shows the example of how the weight updated with the SGD.

$$w_{ij} = w_{ij} - \eta.\frac{\partial L}{\partial w_{ij}}$$

With w_{ij} is the weight between 2 neuron in the different layers, L is the loss function and η is learning rate.

In this above equation, there are two main terms, which have to be considered: learning rate η and the derivatives. The partial derivatives can be calculated by the *backpropagation* algorithm. The idea behind the *backpropagation* algorithm is the effective implementation of the chain rule [32]: the partial derivatives are reused. In other words, the information of loss back propagates to all hidden layer's neurons. The loss contribution of a neuron relies on how much it contributes to the original output. The learning rate η is also very important for the learning process. When the learning rate is too small, it could lead to very slow learning, which consumes a lot of time and resources. On the other hand, choosing it too big could lead to the learning process can not convergence (isolation).

Because of the minimal local problem, SGD might not the best approach for a DNN. However, this is the simple example to understand the concept of optimization the parameter of the NN by using backpropagation algorithm.

2.4 Chapter Summary

In this chapter, we discussed the theory of ML, which is essential for the understanding of later chapters. The chapter started with fundamental information on machine learning before DNN and CNN in detail. The structure of CNN is the stacked of multiple layers, each of them has a unique function. The difference in the combinations of the stacked layer leads to different architectures, which suitable for various purposes. At the end of this chapter, we describe how the model of CNN is trained.

3 Semantic Segmentation

In this chapter, we investigate semantic segmentation, a method for semantic scene understanding. In doing so, we target automotive applications to work towards automated driving. In this chapter, we motivate this research topic and discuss the goals of the book at hand. To this end, we start with an introduction of scene understanding and their impact on the automotive field. In section 3.1, we also explain why we choose semantic segmentation method over others. In later sections, we continue with an overview of the significant factors, which affect onto semantic segmentation's performance such as network architecture (Section 3.2), framework (Section 3.3) and dataset for training (Section 3.4). In the end, the summary of this chapter is shown in section 3.5.

3.1 Overview

In reality, we use our senses to gather information about the environment to make effective decisions while driving. In the automotive application, we use various sensors such as camera, Lidar, ultrasonic sensor or sound sensor...to collect the data of the environment. With the improvement of image quality (better sensor technology, resolution, and optics), the visual sensor (camera) become the primary source to gather environment information such as objects, climate or traffic rules. For this reason, the importance of using images for semantic scene understanding increasing rapidly in recent years.

3.1.1 Semantic Scene Understanding

The semantic scene understanding is not only needed to identify the targets but also understand the distribution of targets and the semantic information in a scene. This information is crucial for many automotive applications such as automated driving, path planning or automated parking. In the following, we give brief fundamental information of some main tasks of semantic scene understanding and also their differences with each other.

- Image Classification: Image Classification is the task of scene understanding, which assigns an input image to one label from a defined set of classes. The output is given in term of probability. Despite the simplification of it, Image Classification has a large variety of practical applications. Nowadays, many researchers use CNN to classify image and the most popular dataset for image classification is ImageNet [34]. Figure 3.1 shows the example of Image classification using CNN.

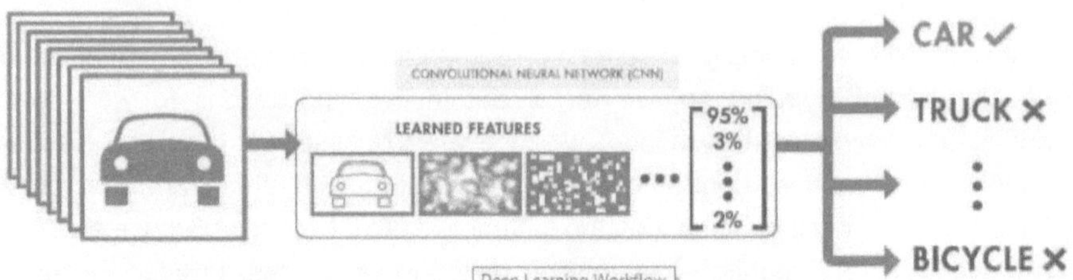

Figure 3.1 Image Classification using CNN [35]

- Object bounding box localization: With object bounding box localization, the task does not only provide the information about the category of a dominant object in the image but also any objects and their localization. The task is outputting bounding (rectangular axially parallel) boxes, labels for individual objects and often confidence score of the label. The most popular application of this task is in autonomous cars where the algorithm not only needs to detect the cars but also any object (pedestrians, motorcycles, trees, building, traffic light..) in the image. The most famous algorithm for object bounding box localization is YOLO (You Only Look Once) [36] [37]. The idea behind YOLO is dividing the input image into multiple grids. After that, it implements both localization and classification algorithm for each grid [36]. The researcher usually uses Microsoft Common Objects in Context (COCO) [38] dataset. Moreover, object bounding box localization is used for other algorithms such as object tracking and multiple object tracking. In multiple object tracking, each object has an ID alongside the label. Online tracking multiple objects is an essential task in automotive which has attracted significant attention. Despite object bounding box localization provide more critical data than image classification, it still has some drawback. First of all, there are some not countable categories such as road, sky, snow, sea and some objects such as buildings or trees are hardly locate by individual bounding boxes. Moreover, object bounding box localization decreases the performance with overlapped or occluded objects.

- Pixel-level semantic labeling: For overcoming limitations of the previous algorithm, we can use semantic segmentation, which describes the process of giving a label for each pixel of an input image with a category label (such as cars, person, trees, trucks). Because of its pixel-accurate category mask, we will use this algorithm for our online perception system. We will go to details of semantic segmentation in the next section. The difference between object bounding box localization and semantic segmentation shows in Figure 3.2.

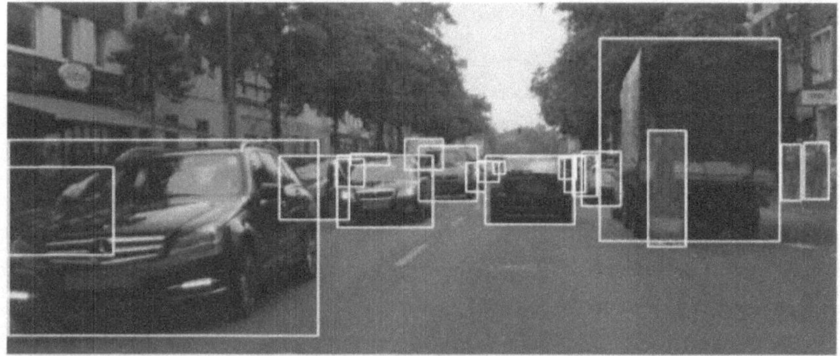

(a) Bounding Box Object Detection

(b) Pixel-level Semantic Labeling

Figure 3.2 Object localization vs Semantic Segmentation [39]

3.1.2 *Semantic Segmentation*

In this section, more detail of segmentation is given, not only about modern methods but also traditional to make clear why we need to use Deep learning or CNN for segmentation nowadays.

Image segmentation task is one of the oldest computer vision tasks, which find groups of pixels which are similar (bottom-up) or because belong to the same object/region (top-down). Usually, this method applies to multiple objects. Semantic image segmentation becomes one of the critical applications in image processing, medical area, automotive and intelligent transportation. There are some outstanding applications using segmentation for detecting road signs [40], detecting brains and tumors [41], land use and land cover classification [42] and in autonomous vehicles [43]. Before the re-rising of DNN (Deep Neural Network), the methods of image segmentation are called traditional methods. Firstly, we give a brief overview of traditional segmentation methods. After that, we focus on the recent progress made by adopting Deep learning.

3.1.2.1 Traditional Segmentation Methods

Traditional approaches for Image segmentation algorithms don't use neural networks and requires heavily in domain knowledge, are wide-spread in the computer vision community.

In the traditional approach, choosing features, which is a piece of information for solving the computational tasks, is very important. Variety of standard features are explained in the following.

- Pixel Color: Pixel color in different image spaces are the most widely used features. RGB and HSL color space are common choices, because of the simplicity (RGB) and illumination invariant (HSL).
- Histogram of Oriented Gradients (HOG): HOG features interprets the image to maps each pixel (x,y) to the gradient.
- Scale-invariant feature transform (SIFT): SIFT feature (global feature) descriptors describe key points in an image.

Next, we go through traditional approaches in image semantic segmentation. The simplest method is thresholding methods, which base on threshold value to turn a gray-scale image into a binary image. This method is widely used in the medical area [44].

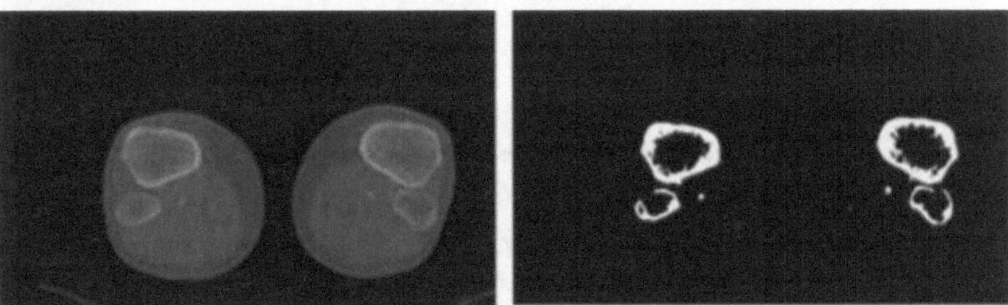

Figure 3.3 Thresholding Segmentation in CT image [44]

Another algorithm is K-means clustering, which is an unsupervised method for clustering. This algorithm is an iterative technique to partition an image into K clusters, which is predefined before the process. Initially, we pick K cluster centers randomly in the feature space. Then it assigns each pixel in the image to the nearest center. After that, it moves the centroid to the center of the cluster and repeats the process until a stopping criterion is reached [45].

Graph-based image segmentation is another method to perform segmentation by describing the image as a graph. In this graph, there are nodes (represented of each pixel), which connect others through edges, which is weighted by the similarity between the two pixels. In this method, image segmentation is performed by finding cuts to divide the image into subgraphs of similar pixels.

Despite the acceptable performance of the traditional method, the main drawbacks of these methods are lacking the ability to interpret the output and semantic information. It leads to the need for semantic segmentation with DNNs.

3.1.2.2 Deep learning

With the increasing availability of large-scale labeled datasets and better computational resources (CPU, GPU), Deep Neural Networks (DNNs) are becoming popular for many semantic scene understanding methods, especially in segmentation. Convolutional networks are driving advances in Deep learning and being used in many research for semantic segmentation [13][14][27]. In semantic segmentation with CNN, we label each pixel with a class. The general idea of using CNN is following.

There is a typical semantic segmentation CNN architecture, which includes an encoder and a decoder network. The encoder produces higher level features via convolution and uses pooling layers to reduce the spatial dimension. Meanwhile, the decoder interprets these higher-level features using class masks and recovers not only the object details but also the spatial information. For more understanding of how semantic segmentation in CNN, we go deep in more specific. Layers in CNN operate convolution to encode the image into higher level representation, which is the combination of a low-level feature such as edges, corners or gradients. These features carry the context of a neighborhood with them. With upsampling and decoding steps, the network decodes these features concerning class associations. Figure 3.4 shows the example for semantic segmentation.

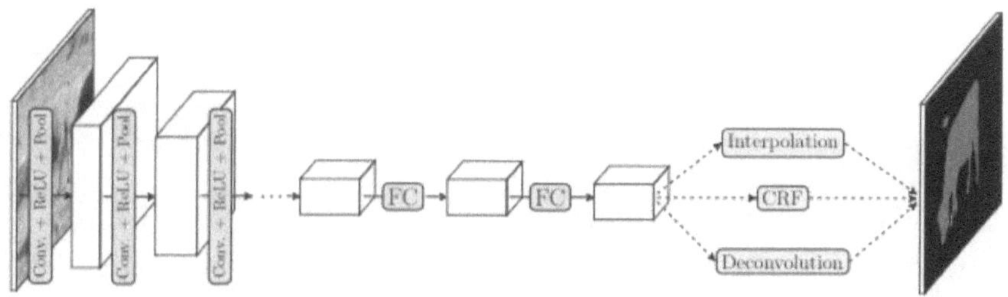

Figure 3.4 Example of Semantic Segmentation using CNN [46]

There are different deep learning approaches. Patch classification is one of the first deep learning approaches. In patch classification approach, we move a patch of the image around to classify each pixel separately [47]. Later in 2014, Fully Convolutional Networks (FCN) [48] is introduced. It becomes a popular CNN architecture for semantic segmentation. The main different of FCN is that it does not use any fully connected layers, only convolutional layers. The main advantage of FCN is much faster than other approach and can work with the different size of the input image. Because of making predictions at a coarse resolution, FCN may have low accuracy in the object boundary region. Another approach is Region-based semantic segmentation (R-CNN). The main difference between R-CNN and FCN is that, in the beginning, the network extract regions from an input image [49]. The different between FCN and R-CNN for semantic segmentation is shown in Figure 3.5.

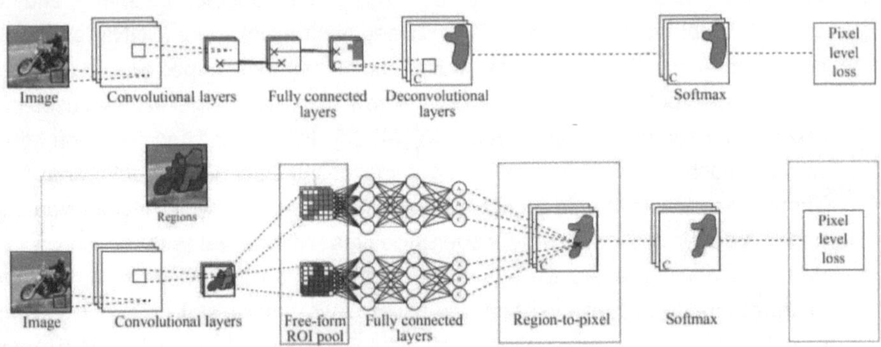

Figure 3.5 The different between FCN (top) and R-CNN (bottom) for semantic segmentation [49]

3.2 Network architecture

For spatially classifying and segmenting images, there are many NN architectures, which have been developed, such as SegNet [13], or many architectures base on FCN [48]. The main idea of these networks is based on a VGG16 [50] architecture. The main drawback of these architectures is a vast model designed with enormous numbers of parameters, which lead to long inference times. With these conditions, they are not suitable for many mobiles, online video processing, automotive, robotic interaction or limited powered applications, which require processing images at rates higher than ten fps.

In this section, we give a theoretical overview of some new neural network architecture, which optimizes for fast inference and decent prediction accuracy.

3.2.1 Efficient neural network

Efficient neural network (ENet) is fast semantic segmentation method [51], which is developed based on ResNets [52]. The structure of ENet is the combination of one master and multiple branches, which separate from the master but will be merged back later. These branches on ENet has convolutional layers. The initial block and the bottleneck module of the ENet are shown in Figure 3.6.

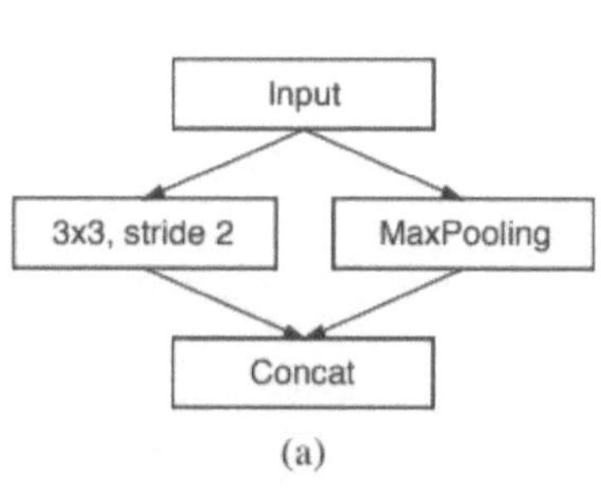

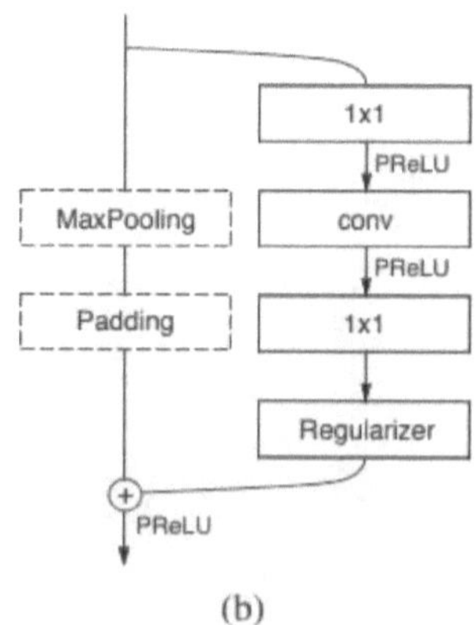

Figure 3.6 ENet initial block (a) and bottleneck module (b) [51]

ENet applies two main approaches to reduce the inference time, which are used in early downsampling and decoder size.

Early downsampling is a crucial step to achieve good performance. We know that processing large input image is very time consuming and expensive. For that reason, ENet implements two first blocks, which strongly reduce the input size but the compressed visual information is still efficient [51].

The main difference between ENet and SegNet is the Decoder size. Instead of using decoders is an exact mirror of the encoder, ENet uses small decoders, which only fine-tuning the details [51].

3.2.2 *Efficient Residual Factorized ConvNet*

Efficient Residual Factorized ConvNet is another approach to achieve top accuracy and efficiency [53], which is fit for real-time application in embedded devices and modern GPUs. ERFNet based on the idea of a new residual block. The main difference of this block is that it uses factorized convolutions.

Such as SegNet and ENet, ERFNet [53] is built with encoder-decoder architecture and the idea of extra layers, which were initially invented in [52]. In [52], there are two designs: non-bottleneck and bottleneck, which are used in many architectures.

Bottleneck model is used for reducing the feature maps to minimize computation (such as ENet in the previous section).

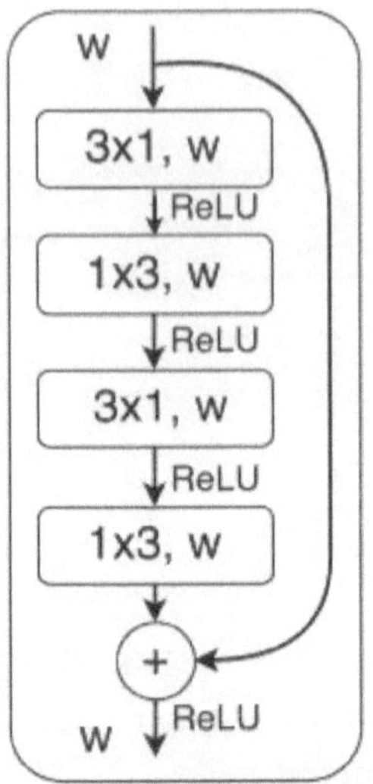

Figure 3.7 Non-bottleneck-1D [53]

In ERFNet, instead of using the bottleneck model, they use non-bottleneck design. The main difference between bottleneck and non-bottleneck is the increase in the internal "width" (feature dimensions). Despite the lower efficiency of non-bottleneck, authors in [53] believe that semantic segmentation tasks gain benefit from additional width. For increasing efficiency, they implement " Non-bottleneck-1D" design, based on the non-bottleneck model, which has the width of the nonbottleneck and also the efficiency of the bottleneck. The "Non-bottleneck-1D" is designed with convolutions one dimension kernels [53]. The main advantages of "Non-bottleneck-1D" are faster execution and reduced number of parameters, without significant impact on its learning performance [53]. The "Non-bottleneck-1D" design is described in Figure 3.7.

3.2.3 Image Cascade Network

Different from previous architectures with using bottleneck and non-bottleneck model [51] [52] [53], image cascade network (ICNet) is a fast segmentation network with decent accuracy [54]. ICNet is the combination of using high inference quality of high-resolution images and low-resolution images, which is cascade image inputs. They produce cascade image input by down-sampling by factors of 2 and 4. The idea of ICNet is described below.

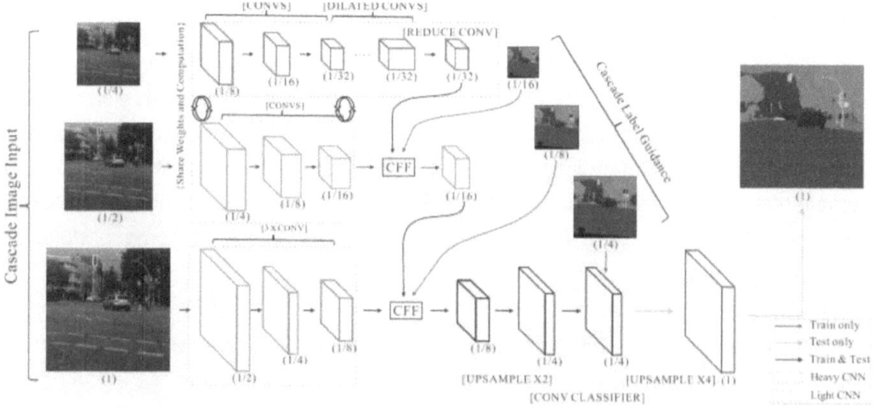

Figure 3.8 Network architecture of ICNet [54]

In the beginning, ICNet uses low-resolution inputs image fed into the heavy CNN for a coarse prediction map. Despite blurred boundaries and missing details of the result, the time to form a coarse prediction map is significantly reduced. For recovering and refining the coarse semantic map, medium and high-resolution features are repeatedly used by light-weighted CNNs[54]. With the multi-resolution branches, the performance of ICNet achieves the decent trade-off between accuracy and efficiency. The comparison between ICNet and other architectures show in Figure 3.9.

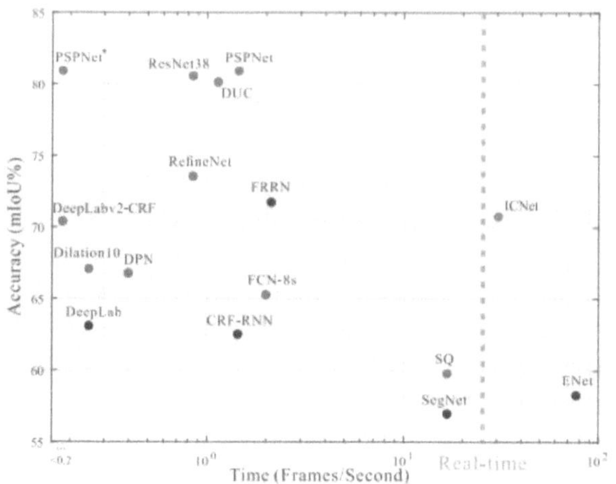

Figure 3.9 Comparision of inference speed and mIoU between ICNet and others [54]

3.3 Framework

In the previous section, some well-known architectures for real-time semantic segmentation are briefly introduced. They are proving that the CNNs can efficiency use for semantic segmentation application, which requires less time-consuming and decent accuracy. But building and deploying applications prove to be a difficult challenge, which affects directly onto time and cost of development of semantic segmentation application. In this section, we have a look at some popular deep learning frameworks.

3.3.1 Torch/PyTorch

Firstly we talk about Torch, which is an open source software library and computing framework. It gives extensive support for deploying and building ML algorithms [55]. For simplification of modifying the existing and designing new machine learning algorithm, Torch defines multiple class as follow:

- **Dataset:** handles multiple types of data such as static or dynamic, memory or hard disk data [55].
- **Machine:** represent a black-box system (neural network, support vector machine, hidden Markov model…), which have parameters and produce output with given input [55].
- **Trainer:** In this class, the set of optimal parameters of a machine is selected and tested by given criterions and dataset [55].
- **Measurer:** use to output several measurements such as classification error, the log-likelihood or the mean-square error [55].

In the beginning, the training example is produced by Dataset and fed forward to Trainer to optimize the parameter of Machine. During the optimization process, the performance of the system can be monitor by Measurer [55].

As different to Torch, PyTorch is implemented on Python, which makes this framework be adopted at a high level. In comparison with Torch, PyTorch framework's architectural is simpler and more transparent [56].

3.3.2 Caffe

The Convolutional Architecture for Fast Feature Embedding (Caffe) provides clean and fully open source machine learning framework, which is supported by many programming languages like C, C++, Python/Numpy, and MATLAB [57]. The Caffe framework becomes popular, because of its speed. With CUDA GPU computation, Caffe can process over 40 million images daily with a single Titan GPU or Nvidia K40 GPU. In other words, one image only needs about 2.5ms to process. Another advantage of Caffe is the separation between the representation and the implementation, which mean the same models can be run GPU or CPU with a variety of hardware (HW) [57]. The highlight of Caffe, which make it different from other framework is given as follow.

- **Modularity:** The software is designed as modular which allows easy extension to new network layers, data formats, and even loss functions [57].
- **Separation of representation and implementation:** Caffe uses Protocol Buffer language to write model definitions as config files. Caffe can easily switch between a CPU and GPU by the implementation of one function call [57].
- **Test coverage:** In Caffe, there is a compulsory that every module needs a test. Moreover, every new code is only allowed to integrate into the project with corresponding tests. This property speed up improvements and refactoring of the codebase [57].
- **Python/MATLAB bindings:** With these bindings, the user can use them to classify inputs and, more important, build networks. It makes Caffe easier to interface with more existing and modern research code [57].
- **Pre-trained reference models:** Caffe provides many pre-trained networks. Users can use these network immediately [57].

3.3.3 Tensorflow

Many researchers consider Tensorflow (TF) [58] as one of the best and most popular deep learning frameworks, which was an open source software library for data-based programming and developed by Google in 2015. The idea of tensorflow is using dataflow graphs (Figure 3.10) to represent state, computation, and operations [58]. The nodes of a dataflow graph are mapped across many machines in a cluster, and within a machine across multiple computing devices such as multi-core CPUs and GPUs. This architecture increases the flexibility of development of the application with a focus on training and inference on deep neural networks [58]. The flexibility of Tensorflow also depend on the following properties

- **Dataflow graphs of primitive operators**: As we mention above, TensorFlow uses a dataflow representation for models. Moreover, individual mathematical operators (convolution, matrix multiplication,etc.) and mutable state (included update operations) are represented as nodes in the dataflow graph, which make the user easily implement new layers with high-level scripting interface [58].
- **Deferred execution**: The execution of Tensorlfow is deferred until the entire program is available, which make the ability for optimization of execution phase by using global information. For achieving this property, the application of TF split into two steps: one for program definition and the second for execution. Despite the efficient performance of this design, there is a need for more complex features into the dataflow graph [58].

- **Common abstraction for heterogeneous accelerators:** Beside general-purpose accelerators devices (multicore CPU, GPUs), there are special-purpose accelerators for deep learning (e.g., Tensor Processing Unit from Google). Tensorflow defines a common abstraction for devices to support all these accelerators, which make the application easily target TPUs, GPUs or mobile CPUs. To make all devices understand common interchange format, TF introduce tensors of primitive values [58].

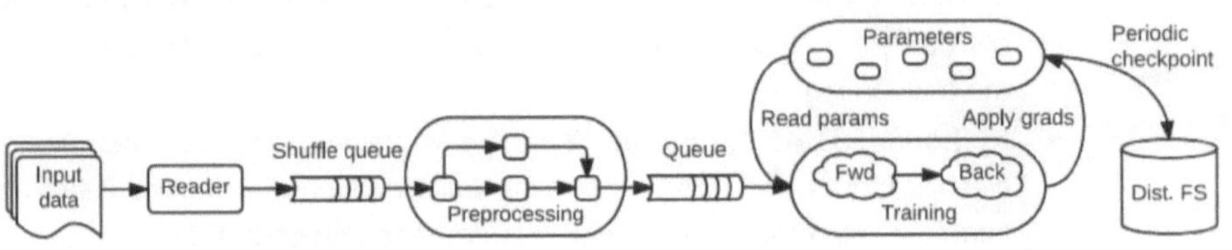

Figure 3.10 A schematic TensorFlow dataflow graph for a training pipeline [58]

3.4   Dataset

The foundation of a machine learning application is the dataset, which not only is used to teach the machine but also testing and evaluation. The dataset effect profoundly onto the performance of the program. Theoretically, the general dataset for ML is spilt into three subtasks:

- **Training set**: A subset data is used to build the network
- **Validation set:** A subset of the dataset, which is used to evaluate the model. This subset is used for fine-tuning model's parameters and selecting the suitable model with given quality standard
- **Test set:** a subset of the dataset to predict the future performance of a model

In this section, we give overviews of some famous dataset, which aim to facilitate the development of semantic segmentation application and method.

3.4.1 Cambridge-driving Labeled Video Database

The video-based database from the University of Cambridge, which is called Cambridge-driving Labeled Video Database (CamVid) is one of the first databases with per-pixel ground truth for 32 classes [59]. CamVid includes elements as following: original video sequences, calibration parameters, camera pose trajectories, and semantic data (list of class labels with pseudo-colors, and hand labeled frames) [59]. Figure 3.11 shows class names and their colors.

Void	Building	Wall	Tree	VegetationMisc
Fence	Sidewalk	ParkingBlock	Column_Pole	TrafficCone
Bridge	SignSymbol	Misc_Text	TrafficLight	Sky
Tunnel	Archway	Road	RoadShoulder	LaneMkgsDriv
LaneMkgsNonDriv	Animal	Pedestrian	Child	CartLuggagePram
Bicyclist	Motorcyclescoot	Car	SUVPickupTruck	Truck_Bus
Train	OtherMoving			

Figure 3.11 List of the 32 object class and corresponding colors in CamVid [59]

The per-pixel dataset of semantic segmentation contains more than 700 images, which was chosen and labeled manually. The reliability of the labeled imaged is improved by requiring the agreement of two-person on each frame [59]. The researcher in [59] also evaluate the benefits of CamVid Database with several existing algorithms (Object recognition, Pedestrian Detection, and Object label propagation) [59].

3.4.2 KITTI dataset

KITTI dataset [60] has been recorded from a VW car, which has many sensor systems such as cameras, laser scanner, and a GPS/IMU. These sensors provide various useful information for research: images, laser scans, IMU accelerations, and GPS data. The dataset has a high diversity, which includes scenes from rural to urban areas with different dynamic and static objects, around Karlsruhe in Germany [60].

These image in KITTI, which is the combination of 8-bit PNG color and grayscale images, divided into five categories: 'Road,' 'City,' 'Residential,' 'Campus' and "Person." The main difference of KITTI with other dataset is that it only label the dynamic object, which classified as 'Car,' 'Van,' 'Truck,' 'Pedestrian,' 'Person (sitting),' 'Cyclist,' 'Tram,' and 'Misc.' Moreover, each object includes its 3D size (height, width, length) [60]. The distribution of object across dataset is shown in Figure 3.12.

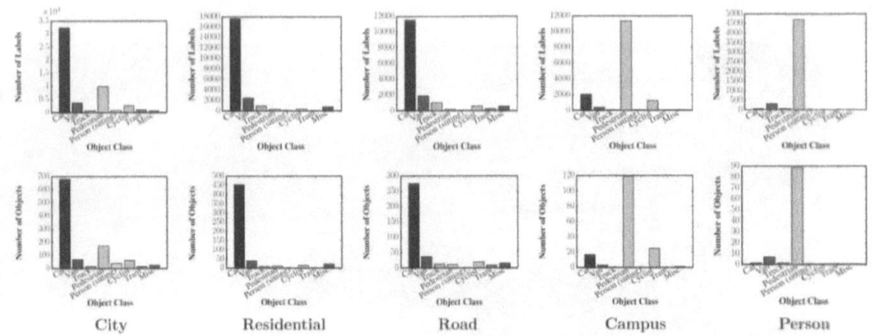

City   Residential   Road   Campus   Person

Figure 3.12 Objects per object class [60]

3.4.3   CityScapes

Cityscape dataset is mainly used for many semantic segmentation applications in an urban environment. The dataset provides 5000 images, which have high-quality pixel-level annotations. Moreover, it also contains 20 000 coarse annotations images. These images are recorded from across 50 cities, mostly in central European, by the 2MP stereo camera, which was installed behind the windshield [61]. The information about the camera specification and location is vital because it has a massive effect on the dataset. In the dataset, there are 30 visual classes, which are divided into eight categories: flat, nature, construction, sky, vehicle, human, object, and void. The distribution of classes over the full dataset is described in Figure 3.13.

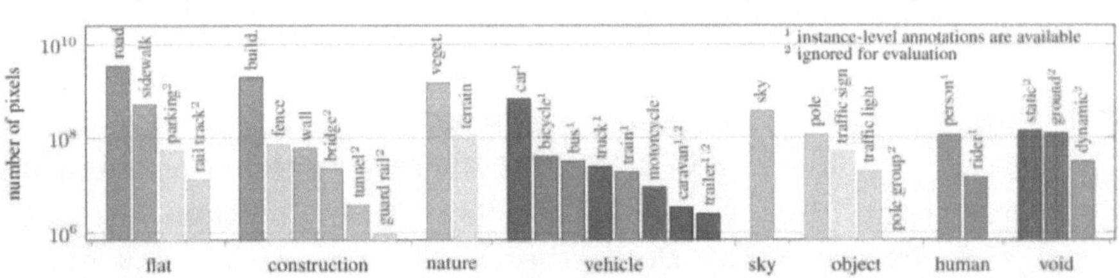

Figure 3.13 Number of finely annotated pixels (y-axis) per class and their associated categories (x-axis) [61]

Cityscape dataset already split into three separate subsets: validation, training, and testing dataset. The images in this dataset are not split by randomly but, images are grouped up for each city, so each subset have different street scene scenarios [61]. The researchers in [61] trained an FCN model with Cityscapes [61], KITTI [62] [63] and CamVid [59] to analyze the performance between three datasets. Table 3.1 shows the result of this comparison.

36

Dataset	Best reported result	CityScapes result
CamVid	62.9	72.6
KITTI	61.6	70.9

Table 3.1 Performance comparison between KITTI, CamVid, and Cityscapes [61]

3.5 Chapter Summary

The development of semantic segmentation application depends heavily on three main factors: network architecture, framework, and dataset. For each factor, the overview of some popular is described. There is no clear answer to which network architecture, framework or dataset is better. The choice of which one depend heavily on the requirement and the resource of each project.

4 Segmentation framework Bonnet

In the previous chapter, we already go through some favourite architecture, framework, and dataset. Despite there are many publication and public information, choosing the right ones is still the critical and difficult decision for the development of semantic segmentation application. For this reason, there is a need for semantic segmentation framework, which allows the user to quickly try new research approaches (framework, architecture, and dataset). The researchers in [1] introduce one of the first frameworks fulfill this requirement, which is named Bonnet.

4.1 Overview

Bonnet, an open source training and deployment framework for semantic segmentation, is firstly introduced by Andres Milioto and Cyrill Stachniss [1]. The idea of Bonnet comes from one of the most significant problems in semantic segmentation field: fragmentation of different systems (framework and architect) and backends (CPU or GPU), which lead to increase the development time and resource. The researchers try to solve this problem by providing a modular implementation tool, which can apply different architecture of CNNs with different backends. With this tool, the users do not need to re-write the whole application, when they want to change backend, device or even CNN. This is the main idea of the creation of Bonnet.

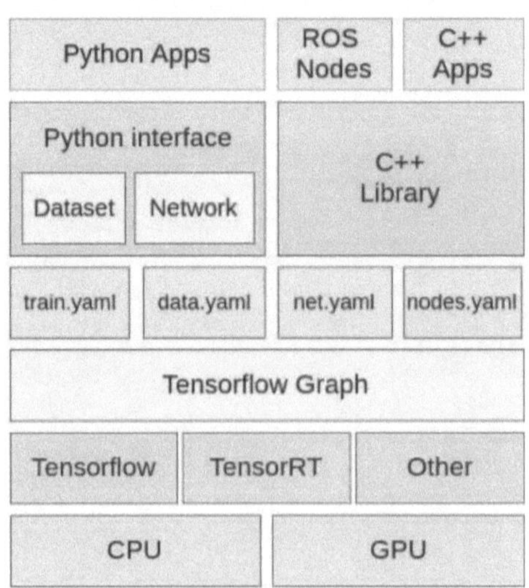

Figure 4.1 Modular description of Bonnet [1]

Bonnet is the combination of a C++ library for a standalone application or ROS-enabled robot deployment and CNN model training pipeline with Python. Figure 4.1 shows a modular description of Bonnet.

4.1.1 Training

At the beginning of using Bonnet, we need to use training pipeline to produce our CNN model for using in the C++ library later. At the beginning of using Bonnet, we need to use training pipeline to produce our CNN model for using in the C++ library later. In Bonnet training pipeline, the architect of CNN and the training dataset are defined with two abstract classes: *Network* and *Dataset*. These two class are responsible for preprocessing data (*Dataset*) and supervised training of CNN (*Network*). The training pipeline is followed below step:

- **Dataset definition**: standardized our data for training, validation, and testing, this step effect the time for pre-processing step.
- **Network definition**: describe how the architect of the CNN looks like, in another word: how to build our CNN.
- **Hyper-parameter tuning**: how to choose hyper-parameter for training. It is the critical section because it effects directly onto the training performance process.
- **GPU training**: is the step, where the training is processed. It could be done with one or multiple GPUs from scratch or a pre-trained model.
- **Graph freezing**: this step freeze the trained model to produce a different optimized model for different supported backends and devices.

4.1.1.1 Dataset Definition

In this step, we need to change our dataset, the directory containing images and label, into the standardized way of Bonnet (pre-processing step), so that *Network* class can use our dataset. Bonnet also provide the general dataset parser as the example for user's data parser [1]. The ideal dataset after parsing should contain training subset, testing subset, and validation subset. Each subset contains an image folder and the corresponding label folder [1]. After parsing dataset (pre-processing step), we need to configure the "data.yaml" file. This configuration file contains much essential information to make the *Network* class can know how to use dataset for training and inference of the model [1]. The data.yaml file must be included with follows information: name and location of the dataset, definition and the debugging-colors for each class, image inference size [1]. The *Dataset* class is not only handler file opening and feeding to CNN but also dataset augmentation (rotations, flips, stretches.)[1].The example of "data.yaml" file is shown in Figure 4.2.

```
1   # dataset cfg file
2   name: "general"
3   data_dir: "/cache/datasets/persons/dataset"
4   buff: True              # if this is true we buffer buff_n images in a fifo
5   buff_nr: 3000           # number of images to keep in fifo (prefetch batch) <-should be bigger than batch size to make sense
6   img_prop:
7     width: 512
8     height: 512
9     depth: 3              # number of channels in original image
10  force_resize: True      # if dataset contains images of different size, it should be True
11  force_remap: False
12  label_map:
13    0: "background"
14    1: "person"
15  color_map: # bgr
16    0: [0, 0, 0]
17    1: [0, 255, 0]
18  label_remap:            # for softmax (it must be an index of the onehot array)
19    0: 0
20    1: 1
```

Figure 4.2 Example of "data.yaml" file [64]

4.1.1.2   *Network Definition and Hyper-parameter Selection*

In this step, we have to define our CNN architecture or using the sample CNN from Bonnet.   For defining the graph, Bonnet provides a library of some common layers. If the user wants to add new layers, they can do it manually by using TensorFlow operations. The class *Network* includes the definition of graph, training and inference methods. There are two important configure file for Network class are "net.yaml " and "train.yaml." The "net.yaml" configuration file contains the information of the model architecture such as the name of CNN; the number of kernels, and layers; and hyperparameters of the network. The optimization method is stored in the "train.yaml", which contain training hyperparameters.

After the successful define the network and the dataset, the next step is selecting the hyper-parameter of the dataset, network, and training method.   These hyper-parameters can be configured by "data.yaml" (e.g., the number of images to be stored in RAM), "net.yaml" (e.g., decays and dropout) and "train.yaml" (e.g., momentums and learning rate for the optimizer, number of epochs, number of GPUs, and the batch size) [1].

After all the definition (dataset, network), hyper-parameter has been chosen, and the promising model is formed, we can start our actual training step with multiple GPUs. Bonnet has the feature that we can train our CNN with the help of multiple GPUs, which reduce the training time significantly [1]. The number of GPUs using in training step can be configured in "train.yaml" file. The example process of multiple training (2 GPU) is shown in Figure 4.3.

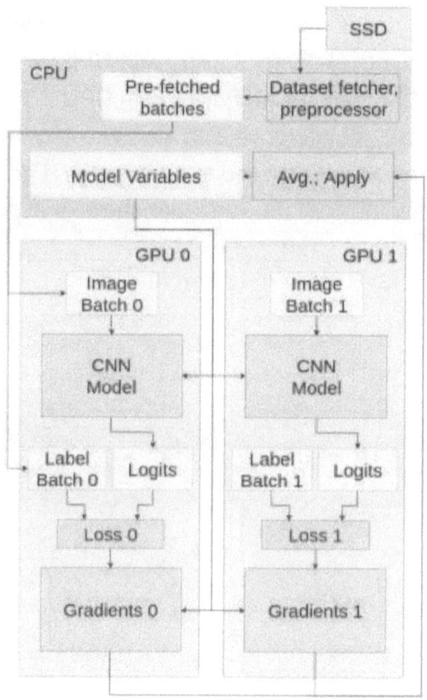

Figure 4.3 Training with 2 GPU [1]

From Fig. 29, we can see that Bonnet performs multiple GPU training by averaging the gradients obtained by a single SGD step in each GPU. GPUs share their model parameters. Bonnet save model parameters in the main memory and update them whenever the averaged gradient is updated. Alongside with all parameters, the accuracy of the network (obtain via validation step) are reported. Moreover, by comparing the accuracy after n epoch, the best model (the best accuracy) is stored by saving the checkpoint method form Tensorflow [1]. Because of this, at the end of training, we can obtain the best model.

4.1.1.3 *Graph Freezing for Deployment*

At the end of the training step, we obtain the output folder, which contains all the configuration file and the best accuracy checkpoint. For using the trained network for deployment, we need the graph freezing step. The freezing step provides the output model, which can be used with different backends (Tensorflow, TensorRT) and devices (CPU, GPU). In this step, all unnecessary information (optimizer ops, the gradients, dropout...) and data are removed. After the freezing step, the output folder provides nodes.yaml (important nodes names) and four frozen models.

- Model in NCHW format: faster if using with GPU
- Model in NHWC format: faster if using with CPU
- Network optimized for inference with the Tensorflow and TensorRT library
- Quantized.pb: network quantized to 8-bit, which is good for mobile.

The example of "train.yaml" and "net.yaml" are shown in Figure 4.4, Figure 4.5.

```
1   # network cfg file
2   name: "bonnet_inception"
3   dropout: 0.25
4   bn_decay: 0.9
5   n_k_lyr: # contains the amount of filters of each layer (res-incept doesn't need it)
6     - 32   # block 1 3x3 downsample
7     - 48   # block 2 3x3 downsample
8     - 64   # block 3 3x3 downsample
9     - 32   # block 1 upsample
10    - 32   # block 2 upsample
11    - 32   # block 3 upsample
12  n_b_lyr: # contains the amount of filter blocks of each dense layer
13    - 1   # block 1 incept down
14    - 2   # block 2 incept down
15    - 4   # block 3 incept down
16    - 1   # block 1 incept up
17    - 1   # block 2 incept up
18    - 1   # block 3 incept up
19  train_lyr:        # boolean list of layers to train, starting by the first conv
20    - True  # block 1 downsample
21    - True  # block 2 downsample
22    - True  # block 3 downsample
23    - True  # godeep
24    - True  # psp
25    - True  # block 1 upsample
26    - True  # block 2 upsample
27    - True  # block 3 upsample
28    - True  # multires
29    - True  # linear classifier
```

Figure 4.4 Example of net.yaml [64]

```
1   # train cfg file
2   max_epochs: 100000
3   loss: "log"           # log (1/ln(fc+e)) or median_freq (median_fc/fc)
4   gamma: 1              # gamma for focal loss
5   lr: 0.0001           # learning rate
6   decay1: 0.9          # decay for first order momentum
7   decay2: 0.999        # decay for second order momentum
8   epsilon: 0.0000001   # epsilon for adam
9   w_decay: 0.00001     # weight decay
10  lr_decay:  200       # decay learning rate every x epochs
11  lr_rate: 2           # decay to 1/x every epoch
12  acc_report_epochs: 10 # every x steps, report accuracy
13  batch_size: 20       # batch size
14  gpus: 4              # number of gpus to use
15  save_imgs: True      # False doesn't save anything, True saves some
16                       # sample images (one per batch of the last calculated batch)
17                       # in log folder
18  summary: False       # verbose summary
19  summary_freq: 50     # steps for summary
20  grads: "speed"       # speed, mem, or tf, for speed optimized, memory optimized, or vanilla tensorflow
21  ignore_crap: True    # last class in cityscapes is crap, therefore we want to ignore it in the metrics
```

Figure 4.5 Example of train.yaml [64]

4.1.2 Deployment

Bonnet help users to quickly deploy a semantic segmentation application by providing a C++ library, which takes care of the inference between the application and the backends. By using the C++ library, the users do not have to worry about backends and device handling [1]. In the beginning, users need to build the network (bonnet class), which need the important information: backend (Tensorflow or TensorRT), execution device (GPU or CPU), and the path to the frozen model (the output of Freezing step in section above). After the successful build of the network, we can use *infer* function, which requires openCV mat as the input and procedure the mask (greyscale image) as the result of the semantic segmentation. Moreover, we can visualize the result of semantic segmentation with the color image (instead of greyscale) by *color* function. The example of how to use Bonnet library is shown in Figure 4.6 [1].

```cpp
#include <bonnet.hpp>
#include <opencv2/core/core.hpp>
#include <string>

int main() {
    // path to frozen dir
    std::string path = "/path/frozen/pb";
    // tf for Tensorflow, trt for TensorRT
    std::string backend = "trt";
    // gpu or cpu (or specialized)
    std::string dev = "/gpu:0";

    // Create the network
    bonnet::Bonnet net(path, backend, dev);
    // Infer image from disk
    cv::Mat image, mask, mask_color;
    image = cv::imread("/path/to/image");
    net.infer(image, mask);
    // If necessary, colorize (like Fig.1)
    net.color(mask, mask_color);
}
```

Figure 4.6 Example of How to Use Bonnet Library [1]

4.2 Advantages

In the previous section, we already go through the fundamental part of Bonnet, how we can use the Bonnet for training and deployment semantic segmentation model. In this section, the advantages of using Bonnet is introduced.

- **Open source framework for deploying and training semantic segmentation network**: This is the first and also one of the most important properties of Bonnet. At this time, if the researchers want to build a semantic segmentation application, they will use some popular framework for machine learning such as Tensorflow, Pytorch or Caffe. The advantages of these frameworks are that they are the multiple-purpose frameworks, which mean we can use it for any application such as object detection, hand-write detection and more. However, the drawback is that they are too general, the user needs some time to get used to it and learn how to use it for their purpose. On the other hand, Bonnet is the open source framework, which is only used for semantic segmentation. Despite the lack of flexibility, Bonnet provides many specific features for semantic segmentation application, which make it more familiar and comfortable to use for the user.

- **Multiple network architectures**: The network architecture is essential for semantic segmentation application. The research needs to choose the one, which one fits with their purposes in term of accuracy and speed. With the help of Network abstract class and a set of common layers, Bonnet allows user can add new or modify the old networks.

- **Multiple dataset:** Such as network architecture, Dataset is a critical factor for the performance of semantic segmentation application. In Bonnet, Dataset is split into three main folders: training, testing, and validation. The main difference between each dataset is the data organization (labels and original images). For this reason, Bonnet provides the standard organization. If we want to add the new dataset for training, the pre-processing step is required.

- **Multiple backends**: Bonnet support multiple Backend. With this feature, the user does not need to rewrite the inference code, when they want to change the backend such as from Tensorflow to TensorRT. They only need to rebuild the bonnet library. This is the excellent feature of the Bonnet. Typically, for changing the background, the developer always needs to rewrite most of the project.

- **CPU and GPU supported:** CPU and GPU support is not the new feature for Machine Learning framework. As we can see from other frameworks, changing the execution device is not always straightforward. In Tensorflow, we need to use the different version for CPU and GPU. However, in Bonnet, the transition between using CPU and GPU is effortless. We only need to change the input variable when we build the network (e.g. using the configuration files).

- **Friendly Training process:** The training process is always the most changeling part of any Machine Learning application. It requires many time and resources, also a deep understanding of Machine Learning. In Bonnet, they try to simplify the way of training. They help the user for handles the optimization method, the graph definition method, and the inference methods to evaluate the performance. Moreover, if the user needs any different optimizer or metric, they can overload the corresponding method [1]. By providing the ability of multiple GPU training, the time for training is reduced significantly.

- **Ready to use resources:** This feature makes the Bonnet is friendly for newcomers. It provides a range of useful resource such as configuration file, pre-processing methods for many datasets and pre-train models. With these resources, anyone can apply it directly to the semantic segmentation application or use them as an example for new dataset and network.

4.3 Limitations

In the previous section, we already see many advantages of Bonnet. It helps researchers save time on CNN training and implementations. These advantages make semantic segmentation application is easily usable for a large number of users, without non-experts knowledge of CNN. On the other hand, Bonnet still shows some limitation on both training and deployment process.

- **Choosing a maximum of epochs is critical:** At the moment, the training process of Bonnet is stopped by reaching the maximum number of the epoch (defined in "train.yaml" file). Choosing the right number is crucial. If the number is too low, the NN become underfitting. However, if the maximum number of the epoch is too high, the NN become overfitting.

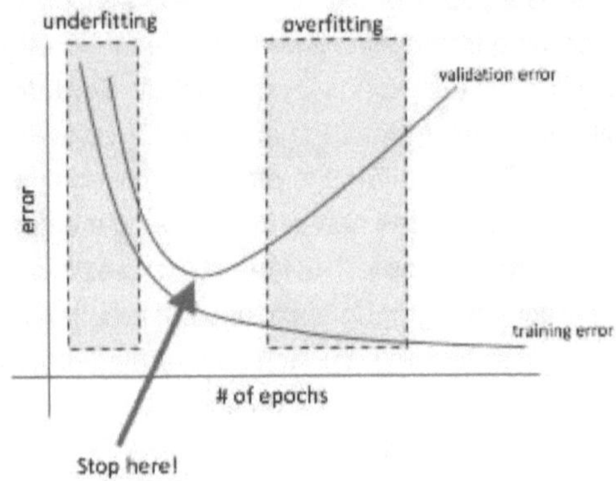

Figure 4.7 Effect of number of epochs [65]

- **Lacking the backup mechanism**: Training a neural network always need much time and computational resource, which depend heavily on the dataset and the complex of the model. Usually, for training a deep NN, it needs about one or two weeks. If we have more computation device (GPUs), we can reduce the time of training. Because of long processing time and heavy computation, we need to take care of any event, which could cause the disruption or termination of the training process (electricity, devices fail-off or human effect). For this reason, the backup mechanism is essential.

- **Lacking testing tool**: During the training process, Bonnet provides the evaluation for validation and testing dataset. However, the stand-alone evaluation tool with the specific dataset is a more convenient way for users. So they can try to evaluate their model without starting the training process, it helps to save much time of development.

- **Only fully supported in Linux**: Despite the aim of Bonnet is "easily usable in robotics and to enable a larger number of people to use CNNs for semantic segmentation on their robots" [1], the Bonnet only fully supported in Linux x86. If the user wants to use Bonnet in ARM architect device, they need to use it along with the ROS, which makes the user have less choice to implement Bonnet in their device. Moreover, Window's user is still a large part of the community, but Bonnet does not support these users.
- **Only supported Tensorflow and TensorRT**: At the moment, Bonnet only supports two backends: Tensorflow and TensorRT. This property reduces the choice of the user and the flexibility of the framework.

4.4 Improvement

In section 4.3 and 4.4, we go through the advantage and the limitation of Bonnet framework. Despite its limitations, Bonnet already shows many advantages over some popular framework that is the reason why we chose to use Bonnet as our framework for deploying and deployment semantic segmentation application for online perception in automated driving. In this section, we implement some new features to make Bonnet more suitable for our project. We try to improve the training, deployment process. Moreover, at the end of this section, we introduce a labeling tool, which is used to label custom images.

4.4.1 Training

As we describe in section 4.3, Bonnet has some limitation in the training process. We use the device with one GPU for training. With this device, the training time of our dataset is roughly more than a week (depend on the number of epoch we choosing). Our device is set up in the public area. All of these conditions (one GPU device and public environment) make our training process is vulnerable with external effect (human and accident). For our purpose, we need to make the process to become more reliable. We decided to implement the online backup mechanism and early stopping.

4.4.1.1 Early Stopping

From figure 4.7, when we train a neural network, the number of epoch effect the network performance (underfitting or overfitting). Generally, all standard neural network architectures are prone to overfitting [66]: the error on the training subset dataset decreases but the error on validation dataset increases. For overcome the overfitting problem, early stopping is the simple method to understand and implement [67]. The simple steps to implement early stopping with validation dataset are [67]:

- Provide two completely independent datasets: training and validation.
- Using a training dataset to train CNN and validation dataset to evaluate the performance of CNN in every n-epoch.
- Stop the training process when the performance of CNN has not increased anymore on the validation dataset.

The simple steps above only become effective when the training the validation set error is perfectly like figure 4.7. The chart in figure 4.7 only shows the ideal training process. In reality, it is far more complicated (figure 4.8)

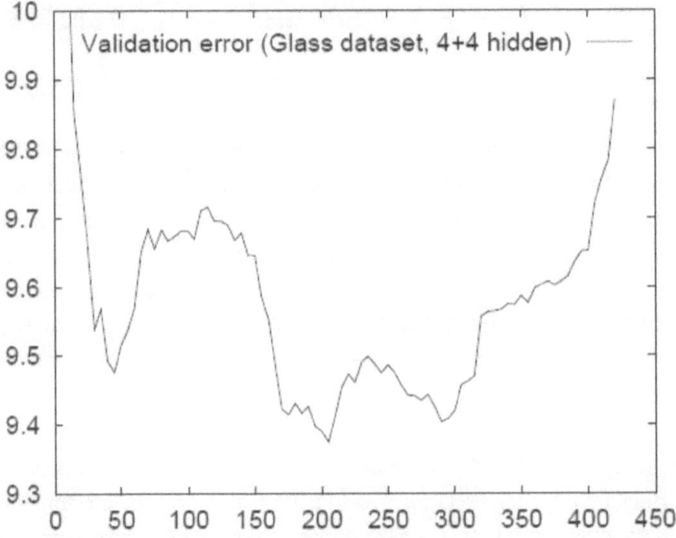

Figure 4.8 A real validation error curve [67]

The main difference between the chart of validation error in figure 4.7 and figure 4.8 is the number of the local minimum. The chart in figure 4.7 only has one local minimum, also global minimum, which makes the simple early stopping method work. On the other hand, the real curve of validation error has many local minima. The simple early stopping method is not reliable anymore.

For this reason, we modify the early stopping method. The idea behind our method is that we do not stop the training intermediately when the error of validation increase. The training process only stops when the performance of CNN is not increasing after a certain number of epochs. Our modifying early stopping method is following these steps:

- Firstly, we start to train CNN with the training dataset.
- After *n*-epoch, we evaluate the performance of the network with the validation dataset. If the performance is improved (highest accuracy), we save the model and accuracy.
- If the performance of CNN is not improved in a certain time, which mean the highest accuracy does not change in a certain number of consecutive epochs.

In this method, the number of consecutive epochs, which model has not improved any more, is critical. The higher this parameter, the more likely the training process can overcome the local minimum and reach the global minimum of a validation error. However, the higher of this parameter also lead to an increase in the time for the training process.   We must define this parameter before the training process. This parameter depends heavily not only onto the size but also the structure of the dataset.

4.4.1.2 Online Backup Mechanism

Training a CNN model can take several days [68]. Because of the availability of the resources (e.g., CPU, GPU, power supply), a long-running training job is likely to experience failure or accidentally termination. For this reason, we need some form of back-up mechanism for fault tolerance. Bonnet use the checkpoint method from Tensorflow for their saving and restoring method, which only saves the best accuracy in the local driver.

In this method, Tensorflow saves their graph by two main files:

- Meta graph file (.meta): protocol buffer saves the complete Tensorflow graph (variables, operations, collections….).
- Checkpoint file: Binary files contain values of the biases, weights, gradients and all the other variables.

The procedure of *Saving* and *Restoring* checkpoint with Tensorflow is shown in Figure 4.9 [69].

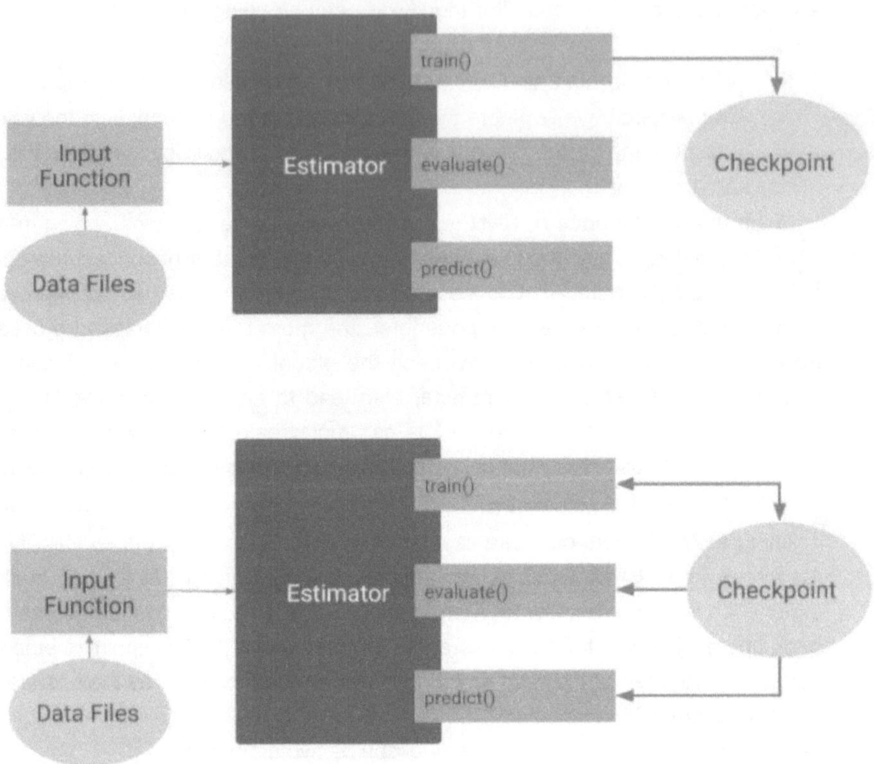

Figure 4.9 Saving and Restoring in TF [69]

Originally, Bonnet only saves the best accuracy checkpoint, and if we want to restore from this checkpoint, we have to give the path to the folder containing the checkpoint manually to the training pipeline. We believe that only saving the best accuracy checkpoint in the local hard drive is not the best solution to handle the resource failure (CPU, GPU, and hard drive). For this reason, we implement our backup and restore mechanism. In our method, during the training process, we save the checkpoint in the external hard drive. Moreover, each checkpoint is contained in a specific location, which is defined by the commit of the codebase. It means that, if we do not change the training pipeline codebase, every time we start the training process, the latest best model is automatically restored and keep training. This feature increases the dependability and autonomous of the training process. If there is some resource failure during the training process, we only need to restart the pipeline, and it will automatically load the latest best model from the external hard drive and keep training the network.

4.4.2 Deployment

In Bonnet, they deploy the C++ library, stand-alone application, and ROS via using Catkin, which is the official build system of ROS. The workflow of Catkin is quite similar to CMake's, but it can make automatic 'find package' and can build multiple, dependent projects. Catkin has many advantages such as the better distribution of packages, better portability, and better cross-compiling support [70].

Despite its advantage, using Catkin not only limits user flexibility, if they want to use Bonnet without ROS; but also makes Bonnet is not available for Window. For this reason, we decide to remove Catkin for building C++ library. Instead of catkin, we use classical CMake's workflow. It increases the flexibility of the software and allows us to distribute Bonnet to Windows and pure version of Bonnet (shared library) to Linux and Arm. This improvement makes users freely use Bonnet C++ Library in many ways, depend on their purpose.

4.4.3 Evaluation-pipeline

As we pointed out in section 4.3, Bonnet does not have a stand-alone evaluation-pipeline, which evaluate the performance of the model without starting the training pipeline. For our project, determine the performance of the model is critical. For this reason, we create the evaluation-pipeline, which based mostly on the training-pipeline of Bonnet. The evaluation-pipeline need the dataset (original image and the ground truth) and the freezing graph folder (from the last step of the training model of Bonnet in section 4.1.1.4) as inputs. The evaluation-pipeline will restore the model and do the semantic segmentation to all image in the dataset. After that, it uses the ground truth images to evaluate the performance of the model. The example of the output of the evaluation-pipeline is shown in Figure 4.10.

```
Test data
   Num samples: 73, Time to run 13.312 sec (only inference: 4.132 sec)
   Network FPS: 17.665
   Time per image: 0.057 s
 Pixelwise Performance:
   Mean Accuracy: 0.6396, Mean IoU: 0.2317
   Intersection over union:
     class 0 IoU: 0.360681
     class 1 IoU: 0.213480
     class 2 IoU: 0.344405
     class 3 IoU: 0.010354
     class 4 IoU: 0.148054
     class 5 IoU: 0.326451
     class 6 IoU: 0.327062
     class 7 IoU: 0.401697
     class 8 IoU: 0.652326
     class 9 IoU: 0.000000
     class 10 IoU: 0.901212
     class 11 IoU: 0.154181
     class 12 IoU: 0.001039
     class 13 IoU: 0.485401
     class 14 IoU: 0.084079
     class 15 IoU: 0.165380
     class 16 IoU: 0.000000
     class 17 IoU: 0.000000
     class 18 IoU: 0.057305
     class 19 IoU: 0.000000
   Precision:
     class 0 Precision: 0.364046
     class 1 Precision: 0.341603
     class 2 Precision: 0.366584
     class 3 Precision: 0.012382
     class 4 Precision: 0.319453
     class 5 Precision: 0.488106
     class 6 Precision: 0.523516
     class 7 Precision: 0.477090
     class 8 Precision: 0.874504
     class 9 Precision: 0.000000
     class 10 Precision: 0.989053
     class 11 Precision: 0.219960
     class 12 Precision: 0.004232
     class 13 Precision: 0.512316
     class 14 Precision: 0.131064
     class 15 Precision: 0.808528
     class 16 Precision: 0.000000
     class 17 Precision: 0.000000
     class 18 Precision: 0.057824
     class 19 Precision: nan
```

Figure 4.10 Example of Evaluation Pipeline output

4.4.4 Image Labeling Tool

The image datasets with ground truth labels are significant for any ML method of object categories, especially in semantic segmentation. In section 3.4, we already discussed the importance of dataset onto the semantic segmentation application. Many researchers worked in this field to create the dataset for semantic segmentation, such as Cityscapes [61].

Despite the fact that building a dataset with corresponding ground truth is a costly and time-consuming process, we still need to build our image dataset with ground truth labels to evaluate the real performance of our system. For this reason, we implemented a new simple image labeling tool, which is designed to fit with our dataset. In this section, we go through some specification (how to use it) and some example result of this tool. The graphic user interface (GUI) of the tool is shown in figure 4.11.

Figure 4.11 GUI of labeling tool

For simplification and saving time, we do not use the original image as the input of the labeling tool. We use the pre-segmented images, which are segmented by traditional image segmentation method, which use standard operators like minimum spanning tree. Figure 4.12 below shows the workflow of our labeling process.

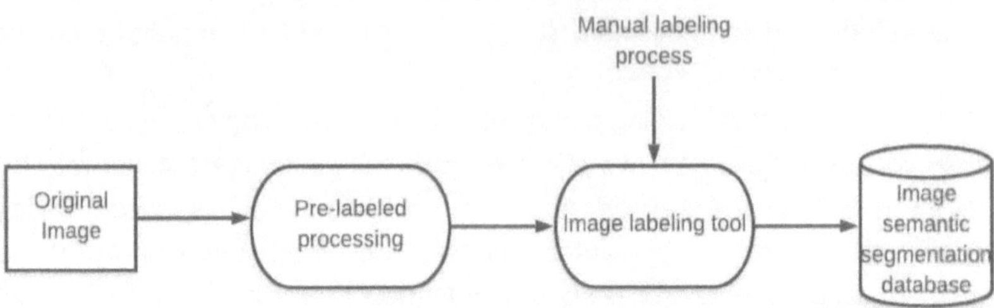

Figure 4.12 Labeling workflow

After the original images are segmented, we use our labeling tool to label the image. We can see from the labeling tool GUI in figure 4.11, the left panel of the tool shows the available labels, which is arranged by category; with the corresponding colors. In the middle, the processing image is placed. On the right side, there are many tools to help us correctly label the image.

The most of these tool in the labeling software is very likely the tools from Paint, which includes brush; drawing line, rectangle and circle; and eraser. There are two special tools: "Flood fill" and "Hole fill", which are based on OpenCV function. Flood fill function is used to color connected pixel (which have the same value or almost the same) with the given color. The threshold of flood fill function is modifiable in the middle of the right panel. The Hole fill is used in a different purpose, which fills the hole inside the object. For quickly and correctly labeling, we can choose the background, which blends (control by alpha parameter) over with label image.

Moreover, we could use the "Correct label" function to change the color of a whole object (not for individual pixel). The result of labeling image is saved as a png file, which has the same name with the original image and had the prefix "_label". If there is any error during the labeling process and we want to correct the label image, we only need to open the original image; the labeling software automatically loads the pre-labeled and the label image.

Figure 4.13 Example result of labeling process

4.5 Chapter Summary

In this chapter, we presented Bonnet, a framework for semantic segmentation; which we use our system. Bonnet has many advantages such as open source framework for semantic segmentation; support multiple network architectures, backends, datasets and execution devices; friendly training process and ready-to-use resources for training and deployment. These advantages make Bonnet become the promise framework for research purpose. However, Bonnet still has some limitation (lack of failure tolerance in the training process and not supported Window). These limitations lead to the need for improving Bonnet to fit and suitable for our project. We not only improve and add some new features for training and deployment process but also introduce a new simple labeling tool. The labeling tool is used to create our dataset, which could be used to improve our work in the near future.

5 Implementation of Online Perception System with *Link*

In the online cooperative perception system, each vehicle needs two crucial abilities, which are sensing and communication. The sensing is the ability to understand the environment, and communication is the ability to share knowledge with others. The idea of the system is that each vehicle uses their sensor system to acquire the information of data and share these data to create a cocoon of perception (figure 5.1).

In the previous chapter, we already discuss the importance of scene understanding and why we chose semantic segmentation over other methods. Moreover, we also found a suitable framework (Bonnet), which help to train and deployment the semantic segmentation application. With this knowledge, we start to implement semantic segmentation into the research vehicle.

At the beginning of this chapter, we describe the development environment of the system. After that, we explain the communication framework in the system. The completed design of the system is given at the end of this chapter.

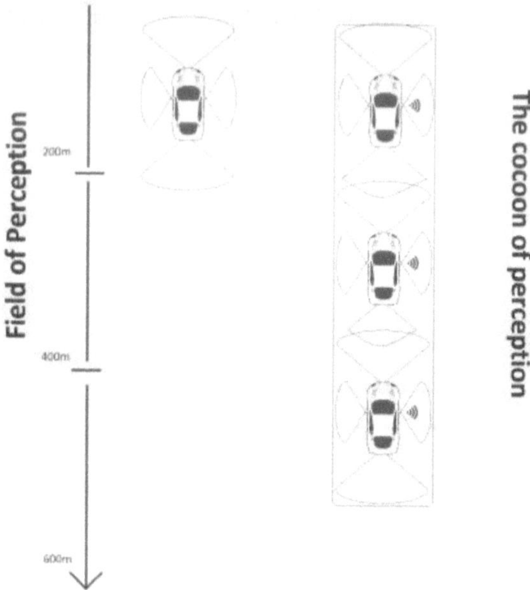

Figure 5.1 Cooperative Perception System

5.1 Development Environment

5.1.1 NVIDIA Drive PX2

The main aim of the autonomous car is creating a safer, more efficient driving environment for the driver; and reduce the cost of transportation. To realize these revolutionary ideas, the autonomous vehicle requires a massive amount of computational power. For overcoming this problem, we need both hardware and software solutions. With the rapid development of deep learning in the automotive domain, there is the need for the powerful and compact computer-in-car that's not only ready to process the huge amount of data but also is compatible and supported with deep learning.

NVIDIA DRIVE PX2 is a lunchbox-sized AI supercomputer for self-driving cars, which is designed to help cars navigate, avoid obstacles and recognize signals, signs, and lanes. PX2 support many input data from different sensors such as cameras, radar, lidar, and other sensors. With the top performance of 24 Deep Learning – Tera-Operations Per Second (DL TOPS), 24000 billion deep learning operations per second, PX2 can handle heavy deep learning process. Moreover, Nvidia also provides a development tool for programmers, which is called DriveWorks. The DriveWorks SDK provide not only the C++ API for programming PX2 but also many examples, which help users know how to use the API to access sensor data and process these data with Deep NN [70]. The specification of the PX2 AutoChauffeur is shown in Table 5.1.

SoCs	2x Tegra "Parker"
CPU	12 Core CPU
CPU Architecture	8xA57 4 x Denver2
GPU Architecture	Pascal (256 Core)
GPU	2 Pascal dGPU, 2 Pascal iGPU
System Memory	8 GB LPDDR4

Table 5.1 Specification of PX2 [70]

Figure 5.2 NVIDIA DRIVE PX2 AutoChauffeur

5.1.2 Gigabit Multimedia Serial Link Camera

Images provide most of the essential data for the driver but are also suitable as an input parameter for many autonomous functions. In our system, the sensing (semantic segmentation) task uses visual data, which rely heavily on the quality of cameras. We need a camera that has high quality (resolution and field of view), but also compatible with our computer-in-car (PX2). For this reason, CMOS Image Sensor AR0231 is the visual sensor of our system.

	Parameter	Value
LENS	Model	SEKONIX NV101
	Construction	1GM5G
	View angle	H : 100° , V : 60°
Sensor	Model	AR0231 (ONSEMI)
	Pixel output interfaces	14-bit parallel
	Input clock	27Mhz
Interface	LVDS(POC)	MAX96705 (MAXIM)
	Connector	FAKRA Z TYPE
Module	Data Interface	LVDS 12bit
	SIZE	30 X 30 X 24.2mm (without optic & cable)
	Power	8V
	Optical Axis	± 5 pixel
	Water Proof	TBD
	Operation Temp	-40 ~ 85 ℃
	Storage Temp	-40 ~ 105℃

Figure 5.3 AR023 specification summary [71]

5.2 TensorRT

In section 3.3, we already discussed some popular framework such as Torch, Caffee, and Tensorflow. Each of them has its advantages and disadvantages, but none of these frameworks are officially installed and supported in NVIDIA DRIVE PX2. Instead of these frameworks, PX2 is fully supported by TensorRT, which optimizes NN models and speeds up for inference across GPU-accelerated platforms (Jetson, Tesla, DRIVE) [72].

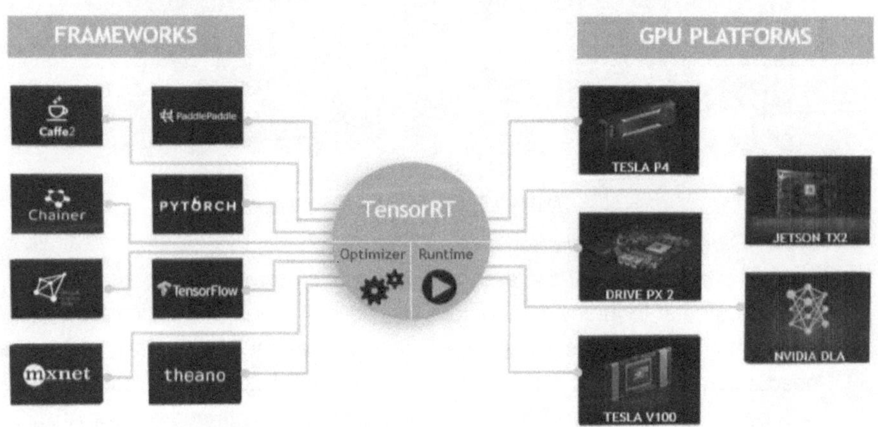

Figure 5.4 TensorRT is programmable inference accelerator [72]

The TensorRT focus on improvement of the performance of a trained network on GPUs. For this purpose, TensorRT provides many parsers for importing existing models from TensorFlow, Caffe, and ONNX. Moreover, users can use its C++ and Python APIs for building the NN application. For optimization NN, TensorRT combines layers and optimize kernel selection for improved latency, memory and power efficiency. Furthermore, if the device does not have large memory, TensorRT can optimize the network to run in lower precision [72].

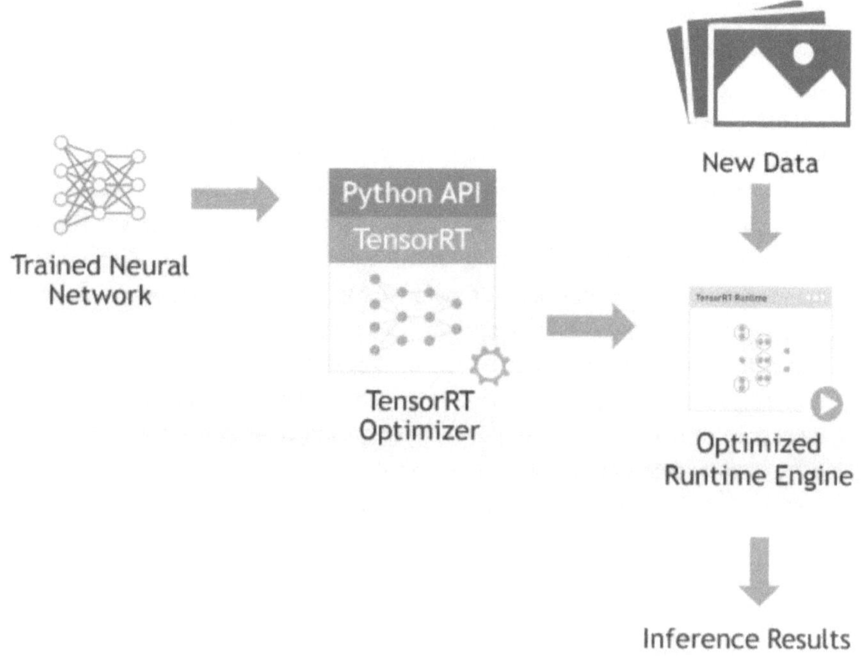

Figure 5.5 TensorRT workflow for deployment and inference [73]

The TensorRT's users mainly come from deep learning engineers, who are working for building application based on new or existing NN model. These applications might be across a wide range of scenarios, such as Robots [74], Autonomous Vehicle [75], and IoT devices [76].

5.3 *Link* - Communication Framework

Communication is one of the most important abilities for autonomous driving in the near future. This ability lets cars broadcast their data such as position, car status (speed, steering-wheel, brake) and other data to other vehicles, system within a close range. The other vehicle can use such information to build a detailed picture of the environment around them. It makes the vehicle could detect the troubles, which are out of the range of individual sensor system.

In this section, we describe the *Link* framework, which is used for communication and real-time data processing [77]. *Link* works with different communication protocols (TCP, UDP, etc.). Main advantages of *Link* are the low-latency data connection and scalable architecture. *Link* also allows building distributed applications in multiple platforms, which are easy to develop and maintain [77]. For understanding the idea behind *Link*, we go through core components of this framework

5.3.1 Node

In *Link*, a system is a mesh of nodes, which are connected and disconnected at any time during the run-time. Each node, which is a small, loosely coupled and reusable piece of software, has a different function (acquiring data, processing data or displaying data) in a system and communicate with each other with pins and messages. One of the best thing about the node concept is that each node is independent with another node. In other words, users can use one node for multiple systems or multiple time in one system. This property saves much time when the user builds their system with *Link*. Based on the role of a node in the communication protocol, we can call a node as a publisher (sending message) or subscriber (receiving message). A node can be the publisher and subscriber at the same time. One node can communicate with multiple nodes, so it has 1 to *n* pins for sending and receiving messages [77]. The example of publish-subscribe architecture is shown in Figure 5.6.

Figure 5.6 Publish-Subscribe architecture [77]

5.3.2 Pin

A pin is a principal object, which helps the node to send and receive messages. Nodes are connected with pins, which act as endpoints (figure 5.6). We have two types of pin: output pin for sending message and input pin for receiving the message. Moreover, each pin also needs to handle subscription management tasks like message ordering, message filtering, message dropping, and bandwidth management. One input pin can connect with multiple output pins, and one output pin can also connect with multiple input pins [77].

- **Output pin**: The output pin works as an encoder, which converts data objects from the node to the messages and sends these messages to another node [77].

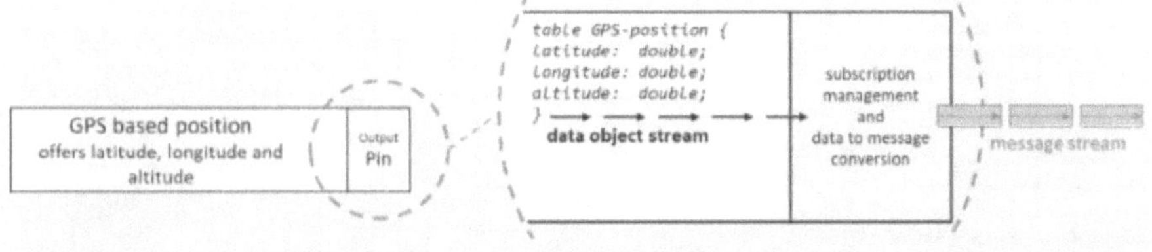

Figure 5.7 Output pin [77]

- **Input pin:** The input pin has the opposite function, which works like a decoder and converts the message back to the data object for the node [77].

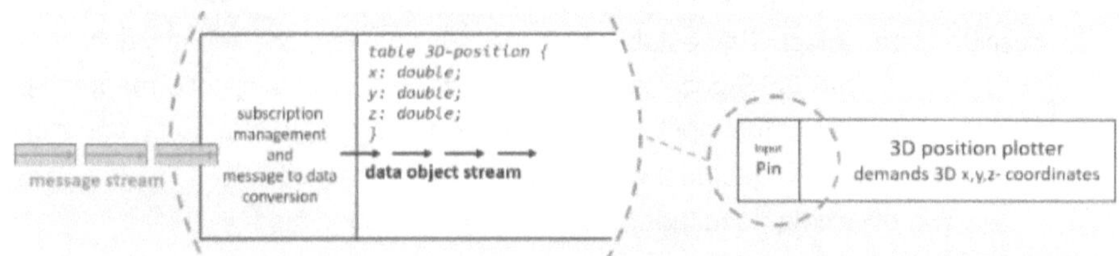

Figure 5.8 Input pin [77]

For correctly connected between the output pin and the input pin, *Link* introduces a concept of *offer* and *demand*. The type of data in the message, which output pin send, is called *offer*. The type of data, which input pin need to receive is called *demand*. The condition for the connection between the output pin and input pin is that the *offer* of output must match with the whole or part of the *demand* of the input pin [77].

5.3.3 Message

In *Link*, Nodes communicate with each other via messages, which can carry different data as payload. Messages are being sent between the nodes with various communication protocols (TCP, UDP, etc.). A flow of messages from the output pint to the input pin is called the message stream. Inside a message stream, each message is transmitted serially and asynchronously [77].

Link uses the concept of a subscription to describe how the demand matches with the whole or part of the offer. When the subscription is established, the subscriber (receiver) can receive the data by mapping, which map data fields of an offer to corresponding data fields of a demand. To configure the mapping process, the users can configure by "instance.json" file [77].

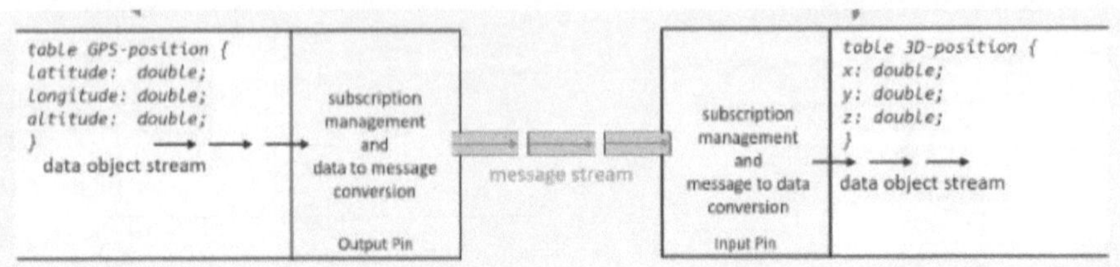

Figure 5.9 Example of Link communication [77]

We look closer to the detail of *Link* communication with an example in Figure 5.9. The mesh in Figure 5.9 contains two nodes. One node has the function of publishing the GPS position data (publisher), and another node is a subscriber, which need 3D-position data object. For establishing the subscription, the input pin of subscriber subscribes to the "GPS-position" offer. During the subscription, the mapping process map the data filed of the input pin demand to the output pin offer in certain ways:

- The offer field "latitude" match with the demand field "x".
- The offer field "longitude" match with the demand field "y".
- The offer field "altitude" t match with the demand field "z".

In other words, "x" receives values of "latitude", y receives values of "longitude", and "y" receives values of "altitude".

5.3.4 Node configuration files

Link provides the way to configure the node by two json configuration files: specification file and instance file. The specification file, which describes the properties of a component, contains the information about pins (output, input) with the corresponding demand and offer, data types in use and user configuration file. The instance of a node is configured by "instance.json" file.

```json
{
    "file-type" : "link2-node-specification-1",
    "$id": "l2spec:/link_dev/ld-node-webcam-2",
    "pins" : {
        "webcam-output" : {
            "pin-type" : "output",
            "offer-type" : {
                "schema-filename" : "./data/Image.bfbs",
                "table-name" : "link_dev.Image"
            }
        }
    },
    "user-configuration-schema": {
        "$schema": "http://json-schema.org/draft-04/schema#",
        "type": "object",
        "properties": { "ImageWidth" : { "type": "number" },
                        "ImageHeight" : { "type": "number" },
                        "FPS" : { "type": "number" },
                        "Grayscale" :{"type": "boolean","default": false}},
        "required": ["ImageWidth","ImageHeight","FPS"],
        "additionalProperties" : false
    }
}
```

a) Specification.json

```json
{
    "file-type" : "link2-node-instance-1",
    "$id": "l2node:/my-mesh/sample-webcam-2",
    "specification": "l2spec:/link_dev/ld-node-webcam-2",
    "pins" : {
        "webcam-output" : {
            "offer-name" : "l2offer:/webcam-output"
        }
    },
    "user-configuration" : {
        "ImageWidth" : 640,
        "ImageHeight": 480,
        "FPS" : 30,
        "Grayscale" : false
    }
}
```

b) Instance.json

Figure 5.10 Example of configuration files of Node in Link

5.4   Online Perception System

In the previous section, we get an overview of the development environment with *Link* communication and data processing framework. We build our online perception system based on this knowledge. In the online perception system, each component is a node and communicated with each other by *Link*.

There are four kinds of component in the system: camera, semantic segmentation, image display, and recorder. A camera node works as a publisher, which is built to work with our camera (Gigabit Multimedia Serial Link Camera). After receiving image data from the camera, a camera node sends the image to the semantic segmentation node. In semantic segmentation node, which works as publisher and subscriber at the same time, the received image data is segmented. The segmented results are sent to other functional nodes such as recorder node and image display node. By using *Link* framework to deploy our components as nodes, each part of the system can be implemented independently, which mean that our system can be built onto multiple platform and devices. We can see these advantages in the next chapter. The complete structure of our system is shown in figure 5.11

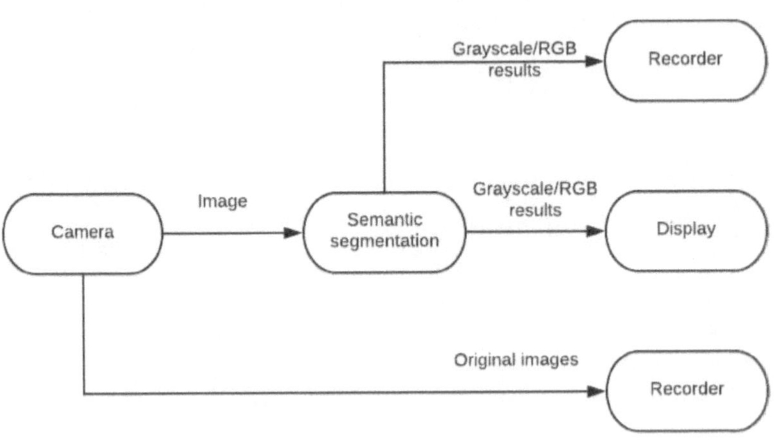

Figure 5.11 Online Perception System

5.4.1 GMSL camera nodes

The digital video cameras work as artificial eyes in an autonomous vehicle. In many autonomous systems, the camera is the primary sensor for acquiring environment information. It also the only sensor that can capture color, texture, and contrast information. At the beginning development stage of online perception system, we only use the camera sensor for gathering image data.

As we know from the development environment, The GMSL camera is used with the NVIDIA PX2 computer-in-car. The NVIDIA PX2 can support to connect up to 12 cameras at the same time. Each node connects to one camera. By using the C++ API from NVIDIA, the node gathers the image from the camera and convert it into Flatbuffer table format, which is called *image*. The *image* data contains the information about the format of the image (grayscale, BGR, RGB or compressed), compressed data and image matrix data.

The GMSL camera node has one output pin, which publishes the images at the same rate with the FPS of the camera. Because PX2 computer can support multiple GMSL cameras, our camera node has the ability to change the camera, which it receives the image, by configuration its "instance.json" file.

When we start the node, it automatically connects to the GMSL camera, displays the property of the camera (FPS, resolution) to the console and sends the image. When the operation of the node is interrupted by the user or camera disconnected, the node displays the summary of the process to the user.

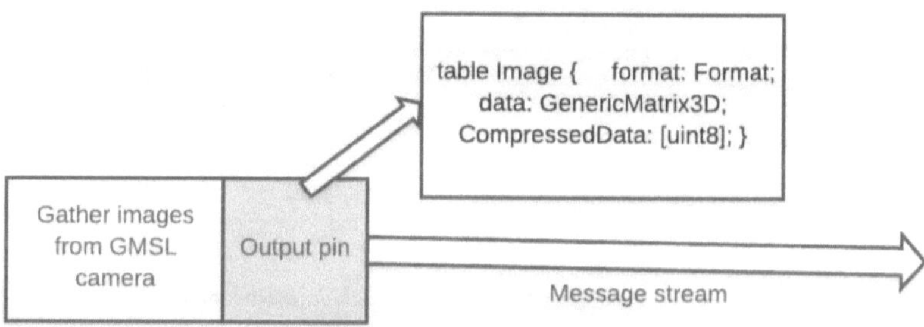

Figure 5.12 GMSL Camera node

5.4.2 Semantic Segmentation Node

In our online perception system for automated driving, we need to understand the scene around the vehicle. As we know from the previous chapter that semantic segmentation method has many advantages over others (chapter 3), we chose this method for our project.

We built the semantic segmentation node by using Bonnet framework. With the help of Bonnet, our node has high flexibility. It can work with difference backends (Tensorflow or TensorRT), CNN models and computational devices (CPU or GPU). The user can configure these properties in the node's "instance.json" file. The semantic segmentation node requires the image data as input, which can be received by the GMSL camera node. At the beginning of the operation, the node loads the information about the backend, the location of CNN model (trained and freeze by Bonnet) and device in configuration file. After that, the node uses this information to build CNN. The time of building CNN is varying and depend on the host device and the backend (Tensorflow or TensorRT). If CNN is successfully formed, the node is ready to receive the image. Otherwise, the node will stop and inform the reason to the user. When the input pin of the node receives the image from the message stream, the node converts the image data into the native OpenCV mat, which can be feed to the CNN. The semantic segmentation process produces two outputs, which are grayscale image and color image. By using two separated outputs pin publish these two results, we can use both results for later processing. The node remains its operation until the user decides to stop the node. In this situation, the semantic segmentation node works as a subscriber (receive image data) and publisher (send the result of semantic segmentation) at the same time.

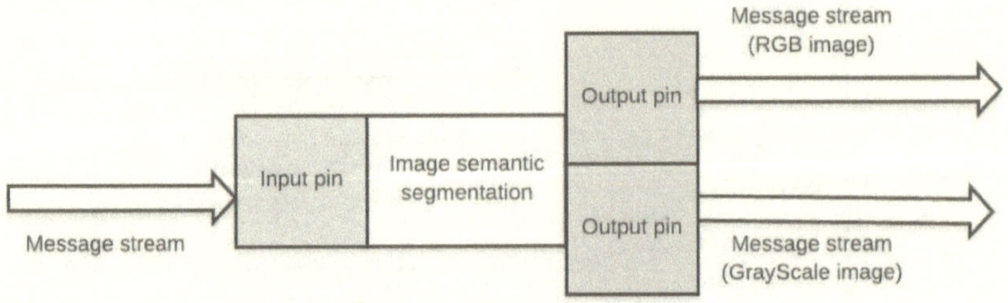

Figure 5.13 Semantic Segmentation node

The difference between the processing rate of semantic segmentation node and camera node lead to the case, where the input image data keep arrives at the node, while the node is busy. This case makes the memory of the PX2 running out, because of the node keep stacking the image into its queue. For preventing this problem, *Link* provides the property "dataobject-queue-capacity," which allow the input pin to drop out the image data object, whenever the node is still processing semantic segmentation an image.

5.4.3 Image Display and Image Writer node

The segmented images are essential data for autonomous driving. For this reason, we implement two functional nodes, one for displaying the result and one for saving these data for offline data processing. The images displaying node and image writer node have one input pin, which receives the image data. After the conversion of image data to OpenCV mat, the images displaying node will display the mat by the help of OpenCV. As we already explained in the previous section about the flexibility of the node in *Link*, the image writer can be used twice in one mesh. One node saves the original image from the camera, and the other one saves the result of semantic segmentation node (figure 5.11). The image writer nodes save the images as the uncompressed png files on to the specific location, which is configurable by "instance.json" file.

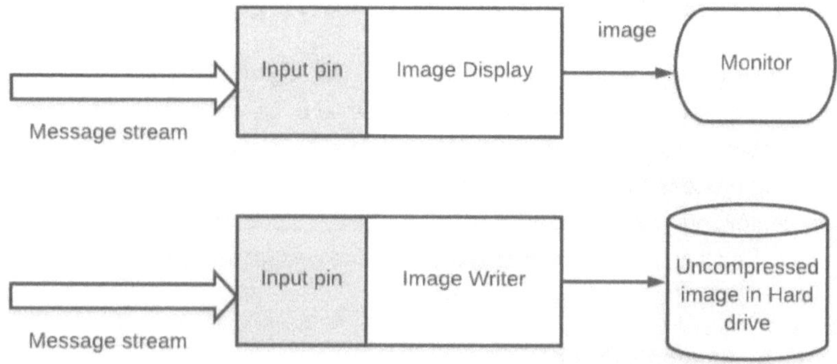

Figure 5.14 Image Display and Image Writer node

5.5 Chapter Summary

In this chapter, we describe the design of our online perception system. In the beginning, we not only give information about the development environment but also the *Link* framework, which we use for component communication and real-time data processing. After that, we apply this knowledge to build our system. The system contains components for acquiring image data, semantic segmentation process, display images and saving data. Each node of the system can be deployed independently and also multiple times. Moreover, the system can work across various architectures. In the next chapter, we bring our design into reality. We can see how the performance of our system.

6 Experiment and Evaluation

We now go to the last part of the book, the experimental evaluation of our online perception system. This chapter is organized as follows. In the beginning, the evaluation criteria are described. We continue by our experiment with the detail of the training process for our CNN semantic segmentation model before we use this model to deploy our system onto NVIDIA DRIVE PX2 computer-in-car. The chapter is completed with a performance analysis of the system implementation.

6.1 Evaluation Criteria

Because of diverse semantic segmentation applications, there is the difficulty of judging whether the semantic segmentation algorithm has made "a good job." Ideally, the performance of the algorithm should be evaluated by the performance of the end application. But this approach is too difficult to realize in reality, especially in graphics applications where the evaluation often requires a user study.

For this reason, in this book. We evaluate the performance of our system based on the performance of the model with the prepared, labeled dataset. There are standard accuracy metrics (precision, recall) and intersection over union (IoU), which help us to judge our model quantitatively. Each measure has its advantage, so we use both at the same time. It makes the user have a better view of how good is our semantic segmentation model.

6.1.1 Standard accuracy metrics

There are three most common metrics used for semantic segmentation: accuracy, precision, and recall [78]. Firstly we need to describe the notation, which is used in the following functions.

Ground truth class	Predicted class	
	Positive	Negative
Positive	True Positive (TP)	False Negative (FN)
Negative	False Positive (FP)	True Negative (TN)

Table 6.1 Class confusion matrix and notation [79]

The ratio of the correctly classified pixels over all available pixels, which is called "accuracy," can be calculated as follows:

$$acc = \frac{TP + TN}{TP + TN + FP + FN}$$

The accuracy is commonly reported for the mean value across all classes as well as each class separately. Other metric is "precision", which is the ratio of true positives over all elements classified as positives. This metric helps us to define the percentage of the discovered class labels are correct.

$$precision = \frac{TP}{TP + FP}$$

The ratio of the true positive pixels over all positive pixels, which is called "recall,". This metric defines the number of correct labeled pixels that are found and can be calculated as follows:

$$recall = \frac{TP}{TP + FN}$$

6.1.2 Intersection over Union

The Intersection over Union metric, also known as the Jaccard similarity coefficient or Jaccard score, is essentially a method to quantify the performance of multi-class semantic segmentation task [80]. With the given testing dataset, the IoU metric measures the number of common pixels between the ground-truth and prediction masks divided by the total number of pixels present across both ground-truth and prediction. The IoU is calculated as follows.

$$IoU = \frac{TP}{TP + FP + FN}$$

As we can see from the functions above, the IoU result is lower than precision and recall. The reason for that, it considers all true positives, false negatives, and false positives. To evaluate the performance of testing datasets, the IoU score of each class is measured separately and, afterward, compute the averaged over all classes.

In our training process, the mean accuracy and the mean IoU are the two crucial metrics. We can choose which model (best mean accuracy or best mean IoU) fits most with our purpose.

6.2 Training Process

Before we start to begin our experiment, we have to prepare our semantic segmentation model, which is used for the online perception system. In the beginning, we tried to use the pre-trained model from Bonnet.

Our system is working for autonomous vehicles, so we have a look onto the pre-trained model of Cityscape. The researches in [1] also show that the Inception architecture has a better performance in comparison with ERFNet and Mobilenets [1]. We use this model with our semantic segmentation node to verify is that the model is suitable for our system or not. Figure 6.1 and Figure 6.2 are the example outputs from the semantic segmentation node. As we can see from Figure 6.1, the node already segments many main objects in the image such as cars, vegetation, and road. However, the output image still contains a lot of mislabeled pixel, especially in the sky and trees. Moreover, the result becomes worst when the objects in the image are small or in the far distance (Figure 6.2).

Figure 6.1 First example output of semantic segmentation node with pre-trained model from Bonnet

Figure 6.2 Second example output of semantic segmentation node with pre-trained model from Bonnet

The accuracy of the node with the current model is not fulfilling our requirement, which leads to the need for retraining the model of Bonnet. There are several ways to achieve this such as choosing a new model, training a new model from scratch or improving the available model by adding more dataset for training. In machine learning technology, the more the data for the training phase is, the better the model. Base on the limited time and resource, we chose to improve the available model by adding more dataset for training.

The performance of semantic segmentation node with Bonnet model is not satisfied with our requirements. The reasonable issue could be lack of the diversity of Cityscape dataset, which use to train the model from Bonnet. One promising solution for this problem is providing the other labeled dataset which has high diversity property. We think that the result of semantic segmentation node could be improved by applying our external labeled dataset. However, starting the training from scratch is unnecessary. For these reasons, training new dataset with the pre-trained model (Cityscape) become the promise solution. With this solution, we not only can combine two datasets (more sample for training, better the result) but also saving much time and computational resource.

Training is the most important, most time and resource consuming phase in machine learning. Our external dataset has some similarity with the Cityscapes, but it still has some critical difference. Because of this reason, we need a preprocessing step to not only arrange the external dataset into the standard organization for training dataset with Bonnet but also fit the labels with the Cityscape dataset.

In the training phase, we will start with the pre-trained model from Bonnet with the additional dataset. After every epoch, we will run the evaluation to find the best model. At the moment, training only runs 100 epochs. If the model does not improve in 10 epochs, the training will stop and freeze model so that we can use this frozen model in semantic segmentation node.

After successfully training, we test the new model with semantic segmentation node to see whether the segmented output images are improved or not. Figure 6.3 shows that the image output improves a lot regarding accurately labeling and shape of the object. Moreover, we run the evaluation-pipeline with a new model with our dataset. Table 6.2 shows the significant improvement of a new model in terms of accuracy and IoU. We decide that, at the moment of development, the accuracy of the model is acceptable. Now we can use the model and semantic segmentation node for further work and will be improved later in the future.

	Accuracy	IoU
New model	0.9121	0.5854
Pre-trained model from Bonnet	0.7657	0.3404

Table 6.2 New model performance

Figure 6.3 Comparison with new (top) and old (middle) model

6.3 Experiment Implementation

Finally, we gather everything we need to implement our online perception system. The system is built based on *Link* framework, and the completed design of the system is described in chapter 5. In the next section, we express our experiment setting with the real vehicle. After that, the performance of our system is evaluated in terms of speed and accuracy.

6.3.1 Experiment setup

We are developing our system for the autonomous driving vehicle. For this reason, our experiment is carry on with the computer-in-car NVIDIA DRIVE PX2, the GMSL camera, and real experimental vehicle. The details and the setup of these components are the following:

- **Experimental Vehicle**: We use *Carai 1* from Professorship of Communications Engineering in TU Chemnitz as our experimental vehicle, which is equipped with camera systems for environment data acquiring. There are many camera sensors such as visible light cameras and near-infrared cameras for low light condition. Moreover, a GPS system provides highly accurate localization information.

Figure 6.4 Experimental Vehicle

- **NVIDIA DRIVE PX2:** The main processing hardware of our experiment is NVIDIA DRIVE PX2. As we already discussed the advantages of PX2 in the previous chapter, this computer is designed to install in the vehicle (*Carai 1*). We set up the PX2 at the back of the *Carai 1* (Figure 6.5), so it can connect to the power supply of the car and have stable standing during the journey.

Figure 6.5 PX2 setup in Carai 1

- **CMOS Image Sensor AR0231:** Our system only uses one GMSL visual sensor, which is CMOS Image Sensor AR0231 from Sekonix. We set up the GMSL camera inside *Carai 1*, on the windshield of the car. From this location, the camera can capture the clear front-view of the *Carai 1*. We can see the set up (top) and the view from the camera (bottom) on figure 6.6 below.

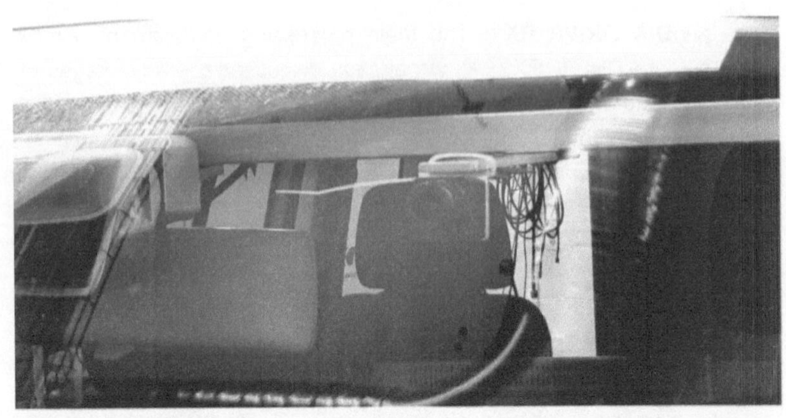

Figure 6.6 GMSL Camera's setup (top) and view from camera (bottom) in Carai 1

- **Trained model:** We know from section 5.2 that PX2 only be supported by TensorRT. However, Bonnet also supports the TensorRT as its backend. Bonnet help us to convert the trained model from Tensorflow to TensorRT and also to inference this model. We do not need to change our codebase of semantic segmentation node to work with TensorRT. After the training and freezing-pipeline by Bonnet. We only need to put the model onto the PX2 and give the location to the semantic segmentation node.

- **Second computer**: As we asserted at the beginning of this book, we are aiming for the cooperative perception system. For this purpose, we set up a second system, which is a laptop to communicate with the PX2. We set up an Ethernet connection between laptop and PX2 so that the PX2 can send the data to the laptop. The laptop not only displays the segmented images but also saves all data to the hard drive. These data can be used for later processed and analyzed. In the future, this laptop can be replaced by another computer-in-car.

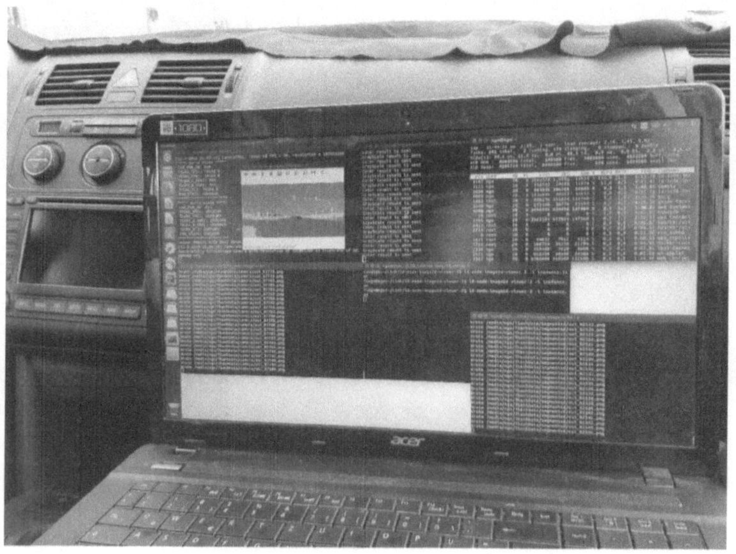

Figure 6.7 The second system

After setting up everything for the experiment, we start our drive test. The drive test is about 15 minutes in the area around TU Chemnitz campus (Reichenhainer Str. 90, 09126 Chemnitz). During the journey, two systems are running in parallel. The PX2 system acquires the images data from the GMSL camera, segments the images and sent all data (original and segmented images) to the second system via the Ethernet connection. The second system, Acer laptop, displays segmented images and saving all data to the hard drive. Because the NVIDIA DRIVE PX2 does not connect with the monitor, the laptop controls every step of the experiment from the beginning to the end with the *SSH* connection. The result of the experiment is shown and discussed in the next section.

6.3.2 Experiment result and evaluation

In this section, we present our experiment result from the test drive. For the cross-validation, we also use the original images from the GMSL camera to do object detection with YOLO [81]. The semantic segmented images (middle) are displayed with the original images from the GMSL camera (top) and the results from object detection method (bottom). In the segmented images, each color represents for each object. The corresponding color of each object is shown in table 6.3. Furthermore, we evaluate our system performance from the test drive with the offline semantic segmentation processing.

Road		Pole		Sky		Bus	
Sidewalk		Trafficlight		Person		Train	
Building		TrafficSign		Rider		Motorcycle	
Wall		Vegetation		Car		Bicycle	
Fence		Terrain		Truck		Crap	

Table 6.3 Color for each object in segmented images

6.3.2.1 Experiment result

Firstly, we begin with one of the most critical situations for autonomous driving, which is the crossroad. When the car come to the crossroad, the driver needs much information about the environment around the car (other vehicles, traffic lights, traffic sign) to make the safety decision.

Figure 6.8 Crossroad

Many researchers use semantic segmentation for localization [82] or 3D reconstruction [83]. These research need knowledge about construction object in the environment such as buildings. Figure 6.9 shows some of these objects.

Figure 6.9 Buildings

Traffic signs are essential information for the driver, and they are also one of the most challenging problems for autonomous vehicles. Localization and understanding the signs are the must for most of the autonomous cars. Figure 6.10 shows how these signs look like in the results of our system.

Figure 6.10 Traffic sign

The autonomous vehicles always take into account the safety of human to the top priority when giving the decision. For this reason, human detection in the image is a critical ability. Figure 6.11 shows how our system can "see" the human in the complex urban scene, which includes cars, traffic lights, traffic sign, and constructions.

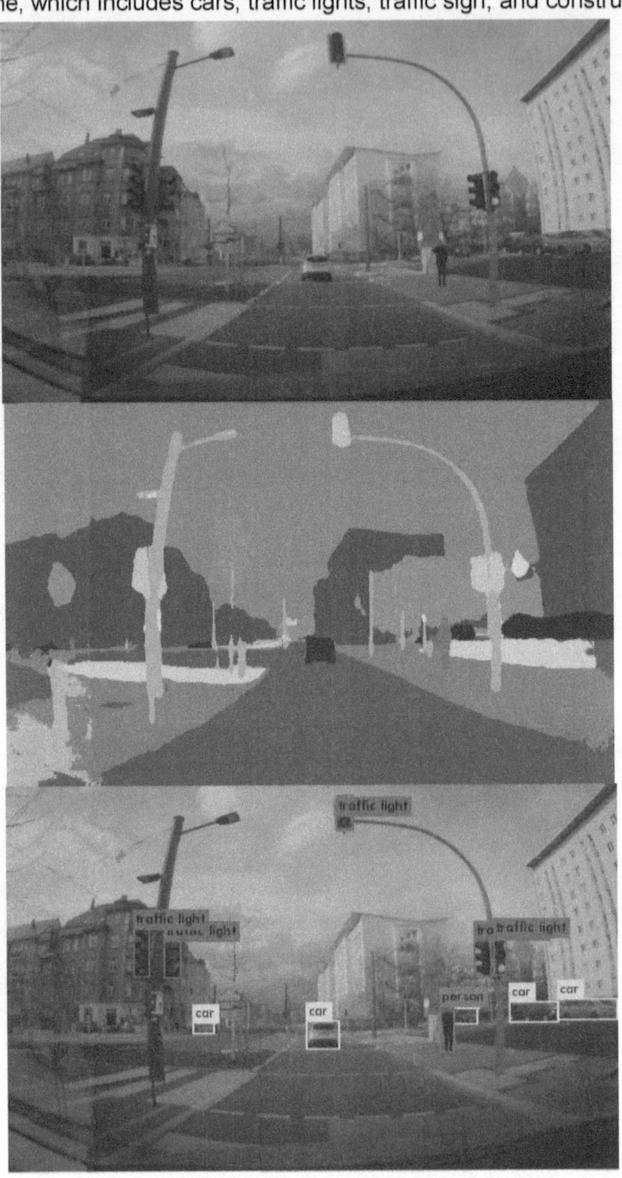

Figure 6.11 Human in complex environment

We can see the results of our experiment from previous figures (6.8, 6.9, 6.10, and 6.11). With the segmented images, we can distinguish many principal objects in the scene such as cars, trees, road, buildings, traffic light, and traffic signs. With the cross-road scene in figure 6.8, the system can detect not only cars from far-away but also important objects such as traffic light and traffic sign. Moreover, the shape of objects in figure 6.8 is also reported. On the other hand, figure 6.10 shows the ability to capture and distinguish between buildings and big traffic sign. In figure 6.11, the complex and popular urban scene is testing with our system. The performance of the system reaches our target. Many important objects are correctly segmented, especially human. We know that the size of the object is effected heavily in any detection task. However, our system can detect the object with small size (figure 6.11 and 6.8).

After the driving test, we use the original images from the GMSL camera to do the offline cross-validation test with object detection using YOLOv3 [83]. From the results of YOLOv3, we can see that our system can detect more object than the YOLOv3. Especially in some static object like buildings, poles, road and environments like the sky and tries.

In many automotive applications, like in image segmentation for autonomous vehicles, the runtime is critical since usually a massive amount of data has to be processed. In this part, the runtime of our system is compared with the offline processing with CPU. And we also examine the runtime between our system with YOLOv3 (only CPU). On the PX2's GPU the system requires around 0.1 seconds to segment one image. Using the CPU (Intel i5-4200M with two cores) lasts approximately 1.5 seconds. Furthermore, for each image, the YOLOv3 needs about 18-20 second to detect objects.

Semantic Segmentation (GPU)	8 fps
Semantic Segmentation (CPU)	1 fps
Object detection with YOLOv3 (CPU)	18 fps

Table 6.4 Runtime evaluation

6.4 Chapter Summary

In this chapter, we evaluate our online perception system with the driving test. Before the driving test, we need to prepare a trained model for CNN. However, the pre-trained model from Bonnet does not meet our requirement for the performance. For this reason, we need to improve the pre-trained model with our labeled dataset. After the training process, the model is enhanced significantly, and we can use this model for our system. In the system, we use NVIDIA DRIVE PX2 as computer-in-car, GMSL camera as a visual sensor and a laptop as the second system. The system is installed onto *Carai 1*, the development vehicle. During the test, PX2 receives the image from the camera, segmented and send all data to the laptop, which is used for displaying segmented images and saving data to the hard drive. From the results of the driving test, we can see that our system has the capability to fully operational in real-time with the acceptable performance.

7 Conclusions and Future Work

In this chapter, we summarize our work to develop the online perception system and give an outlook for future research.

7.1 Conclusions

In this book, we showed that the semantic segmentation method can be successfully implemented for the online perception system. At the beginning of this book, we give an overview of machine learning and semantic segmentation to understand why we use this method for our perception system. Also, we went through some well-known machine learning framework, which could be used for semantic segmentation. After some evaluation, we decide to use Bonnet. Based on the work of Milioto et al. [5], we proposed a modified version of Bonnet semantic segmentation framework to improve the flexibility and the performance. We start to train the model for our semantic segmentation application with the new version of Bonnet. The trained model shows some improvements with the pre-trained model from Bonnet.

After the preparation of the model for semantic segmentation, we move to the main part of this book, which is the designing and implementation of online perception system, which has the ability to cooperate which another system. In our system, communication and real-time data processing are crucial abilities, which are handled by *Link*. In *Link*, each component in the system works as a node, which can send and receive data by messages. For evaluating the performance of our system, in reality, we deployed our design on to the NVIDIA DRIVE PX2 computer-in-car and did the test drive with the development vehicle.

7.2 Future work

From the experimental evaluations, we saw that there is room for improvements. The major challenges are still accuracy and runtime of the system.

To reduce the runtime of the system, we can utilize the maximum power of PX2 by using 2 SOCs (Tegra-A and Tegra-B) at the same time. Tegra-A and Tegra-B are inter-connected with Ethernet over the Marvell Gb AVB Switch. They can communicate with each other through this network connection. By using 2 SOCs, we can decrease our runtime with more memory and more computation devices (GPU and CPU).

Future work could improve the accuracy of semantic segmentation by multiple-view. PX2 support up to twelve GMSL camera at the same time; these cameras can capture images at different views. These data can be used to give a better-segmented image of the environment (wide range of FOV, more accuracy).

Additionally, our labeling tool needs to be improved to reduce huge manual effort to generate pixel-level labeled images. This improvement of the tool can let us create our dataset better, faster and more convenient. The better-labeled dataset, the better model we get.

www.ingramcontent.com/pod-product-compliance
Lightning Source LLC
LaVergne TN
LVHW041738190726
843493LV00008B/2404